NUREG-1854

NRC Staff Guidance for Activities Related to U.S. Department of Energy Waste Determinations

Draft Final Report for Interim Use

Manuscript Completed: August 2007
Date Published: August 2007

Prepared by
H. Arlt, A. Bradford, N. Devaser, D. Esh, M. Fuller, A. Ridge, (NRC)
B. Brient, P. LaPlante, P. Mackin, E. Pearcy (CNWRA)
D. Turner, J. Winterle (CNWRA)

Center for Nuclear Waste Regulatory Analyses
Southwest Research Institute
6220 Culebra Road
San Antonio, TX 78238-5166

Division of Waste Management and
    Environmental Protection
Office of Federal and State Materials and
    Environmental Management Programs
Washington, DC 20555-0001

ABSTRACT

This document provides U.S. Nuclear Regulatory Commission (NRC) staff guidance for conducting activities related to waste determinations. Waste determinations are evaluations performed by the U.S. Department of Energy and are used to assess whether certain wastes resulting from the reprocessing of spent nuclear fuel can be considered low-level waste and managed accordingly. This guidance document applies to NRC activities that may be conducted for the Savannah River Site (SRS) in South Carolina and the Idaho National Laboratory (INL) in Idaho pursuant to the Ronald W. Reagan National Defense Authorization Act for Fiscal Year 2005 (NDAA), as well as the Hanford site in Washington and the West Valley site in New York. The guidance document discusses the background and history of waste determinations, the different applicable criteria and how they are applied and evaluated, the review of associated performance assessments and inadvertent intruder analyses, the removal of highly radioactive radionuclides, and NRC's monitoring activities that will be performed at SRS and INL in accordance with the NDAA.

CONTENTS

CONTENTS (continued)

CONTENTS (continued)

CONTENTS (continued)

CONTENTS (continued)

LIST OF FIGURES

CONTENTS (continued)

ABBREVIATIONS/ACRONYMS

ACNW   Advisory Committee on Nuclear Waste

ADAMS   Agencywide Documents Access and Management System

ALARA   As Low As Is Reasonably Achievable

BTP   Branch Technical Position

CERCLA   Comprehensive Environmental Response, Compensation, and Liability Act

CFR   Code of Federal Regulations

DNFSB   Defense Nuclear Facilities Safety Board

DOE   U.S. Department of Energy

EPA   U.S. Environmental Protection Agency

ET   Evapotranspiration

HLW   High-level Waste

IA   Interagency Agreement

ICRP   International Commission on Radiological Protection

INL   Idaho National Laboratory

K_d Distribution coefficient

LLW   Low-level Waste

MOU   Memorandum of Understanding

NAS   National Academy of Sciences

NDAA   Ronald W. Reagan National Defense Authorization Act for Fiscal Year 2005

NEPA   National Environmental Policy Act

NRC   U.S. Nuclear Regulatory Commission

NRDC   Natural Resources Defense Council

NYSERDA New York State Energy Research and Development Authority

OMB   Office of Management and Budget

PA	Performance Assessment
RAI	Request for Additional Information
RMEI	Reasonably Maximally Exposed Individual
SRM	Staff Requirements Memorandum
SRP	Standard Review Plan
SRS	Savannah River Site
TEDE	Total Effective Dose Equivalent
TER	Technical Evaluation Report
TRU	Transuranic
USGS	U.S. Geological Survey
WIR	Waste Incidental to Reprocessing
WVDP	West Valley Demonstration Project

INTRODUCTION

Foreword

In October 2004, the Ronald W. Reagan National Defense Authorization Act for Fiscal Year 2005 (NDAA) was enacted. Section 3116 of the NDAA gave the U.S. Nuclear Regulatory Commission (NRC) new responsibilities with respect to U.S. Department of Energy (DOE) waste management activities for certain radioactive material resulting from the reprocessing of spent nuclear fuel within the States of South Carolina and Idaho (see Appendix A for the text of Section 3116 of the NDAA). NRC's responsibilities include consulting with DOE in DOE's determination of whether the waste is not high-level waste (HLW), as well as monitoring DOE's disposal actions for these wastes. The concept behind this "incidental" waste is that some radioactive material, resulting from the reprocessing of spent nuclear fuel, does not need to be disposed of as HLW in a geologic repository. This is because the residual radioactive contamination of the material, if properly controlled, is sufficiently low so that it does not represent a hazard to public health and safety. Consequently, incidental waste is not considered to be HLW but instead is low-level waste (LLW) or, in DOE's waste classification scheme, transuranic (TRU) waste. DOE uses technical analyses that are documented in a "waste determination" to evaluate whether waste is incidental, or alternatively, is HLW. A waste determination is DOE's analysis of whether the waste will meet the applicable incidental waste criteria.

Prior to passage of the NDAA, DOE would periodically request that NRC provide technical advice for specific waste determinations. The NRC staff reviewed DOE's waste determinations to assess whether there were sound technical assumptions, analyses, and conclusions with regard to meeting the applicable incidental waste criteria. The staff typically evaluated DOE-submitted information, generated requests for additional information (RAIs), met with DOE representatives to discuss technical questions and issues, and documented the final review results in Technical Evaluation Reports (TERs). NRC's advice was provided in an advisory manner and did not constitute regulatory approval. Similarly, NRC reviews of DOE waste determinations under the NDAA are in a consultative role; NRC does not have regulatory authority over DOE. These types of reviews were completed for waste intended to be removed from tanks at Hanford (NRC, 1997a), tank closure at the Savannah River Site (SRS) (NRC, 2000a), waste intended to be removed from tanks at Idaho National Laboratory (INL) (NRC, 2002a), and tank closure at INL (NRC, 2003a).

Because the NRC expects NDAA enactment to increase the number of waste determinations submitted for NRC review, NRC decided to develop *NRC Staff Guidance for Activities Related to U.S. Department of Energy Waste Determinations* (guidance document or document). This document will provide NRC staff guidance on conducting a technical review of a waste determination, and monitoring activities under the NDAA, and it will help ensure consistency across different reviews and different reviewers. Because the technical aspects of NRC's waste determination reviews are expected to be similar for all four sites regardless of whether the site is covered by the NDAA, this guidance document will address reviews for SRS, INL, the Hanford site, and the West Valley Demonstration Project.

1   In November 2005, NRC held a public scoping meeting in Rockville, Maryland, to obtain public
2   input on the scope of the (then) draft Standard Review Plan (SRP). In addition, NRC published
3   a Federal Register notice on November 2, 2005, announcing the scoping meeting and stating
4   that written comments on the scope of the (then) draft SRP would be accepted until
5   November 25, 2005 (NRC, 2005a). The transcript of the scoping meeting is publicly available
6   in NRC's Agencywide Document Access and Management System (ADAMS) (NRC, 2005b).
7   NRC received written comments on the proposed scope of the SRP from the SRS Citizens
8   Advisory Board (CAB, 2005), the South Carolina Department of Health and Environmental
9   Control (SCDHEC, 2005), and the Washington State Department of Ecology (Washington,
10 2005), and the comments are publicly available in ADAMS.
11
12 In December 2005, the NRC published a Federal Register notice issuing draft interim guidance
13 for performing concentration averaging for waste determinations (NRC, 2005c). The Federal
14 Register notice was issued due to high stakeholder interest in obtaining the guidance as soon
15 as practicable due to the ongoing development of waste determinations. The draft interim
16 guidance was open for public comment until January 31, 2006. NRC received comment letters
17 from the State of Idaho Department of Environmental Quality (Idaho, 2006), the State of
18 Washington Department of Ecology (Washington, 2006), the State of Oregon Department of
19 Energy (Oregon, 2006), Washington Closure Hanford (WCH, 2006), the Natural Resources
20 Defense Council (NRDC, 2006), and members of the public (Greeves and Lieberman, 2006),
21 and the comment letters are publicly available in ADAMS. The concentration averaging
22 guidance that is included in this draft guidance document has been revised from the initial draft
23 document based on stakeholder comments (see Section 3.5.1.1).
24
25 This document was published for interim use and comment in May 2006. NRC accepted public
26 comments on the document until July 31, 2006. NRC received 12 comment letters during the
27 public comment period, which are addressed in this revision. Comments were received from
28 four State agencies, two public interest groups, and five individuals, and the DOE. In addition to
29 the public comments, the staff also received comments and recommendations from the NRC
30 Advisory Committee on Nuclear Waste (ACNW). After consideration of the purpose of
31 NUREG–1854, as well as in response to comments, the NRC has revised the name of the
32 document from a "standard review plan" (SRP) to a "staff guidance document." This revision
33 was made to better reflect the document's purpose. An SRP is typically used by the staff to
34 review licensing actions. Since DOE waste determinations are not licensing actions, the
35 document is being issued as staff guidance. However, the level of detail and the general
36 content of the document remain unchanged.
37
38 NUREG–1854 is a "Draft Final Document for Interim Use" because revisions to the document
39 are anticipated in the fairly near future. NRC, DOE, and the covered States are, at the time of
40 this document's publication, working to resolve a number of generic technical issues (GTI) that
41 will enable all parties to more efficiently develop and review future draft waste determinations.
42 When the GTIs are resolved, this NRC Staff Guidance Document will be updated to include the
43 GTI resolution discussions.
44
45

1   This NRC Staff Guidance Document provides direction to the NRC staff and does not set forth
2   regulatory requirements for NRC nor DOE, and compliance with this review plan is not required.
3   This NRC Staff Guidance Document will be revised and updated periodically to clarify the
4   content or incorporate modifications as needed.
5
6
7
8
9
10
11   Larry W. Camper, Director
12   Division of Waste Management and
13   Environmental Protection
14   Office of Federal and State
15   Materials and Environmental
16   Management Programs

How To Use This Staff Guidance Document

This document provides NRC staff guidance for evaluating DOE-developed waste determinations for SRS, INL, Hanford, and West Valley. The guidance document provides elements the staff should review to determine whether there is reasonable assurance that the appropriate criteria can be met. This guidance document also provides information about the NRC's role in the waste determination process and NRC's monitoring activities under the NDAA.

This document does not set forth regulatory requirements for NRC nor DOE, and compliance with this guidance document is not required. Methods and approaches different from those described here could be acceptable to demonstrate that there is reasonable assurance that the appropriate criteria can be met. Waste determinations typically use the performance objectives of 10 CFR Part 61, Subpart C, as a criterion that must be met (see Section 2); references to other parts of the regulations in 10 CFR Part 61 (i.e., other than Subpart C) are included only to provide information and guidance as they relate to the staff reviews.

Because waste determinations are based on meeting the performance objectives of 10 CFR Part 61, Subpart C, or comparable performance objectives, this guidance document can help reviewers assess whether there is reasonable assurance that public health and safety can be protected as described in 10 CFR Part 61, Subpart C. The term "reasonable assurance" is the standard specified in the general requirement of 10 CFR Part 61, Subpart C (§61.40). In the context of waste disposal activities, NRC staff develops reasonable assurance by (1) evaluating the technical bases for the predicted performance of a disposal facility and resulting potential doses to hypothetical receptors, (2) evaluating sources of uncertainty and conservatism in supporting analyses, and (3) determining whether, given the uncertainties, the conclusions of the analysis are sound. In the review, the staff often performs independent analyses to identify important assumptions in DOE's analysis so that it can assess the degree to which the staff agrees with DOE's assumptions. NRC has previously indicated (NRC, 2001) the term "reasonable assurance" is not meant to indicate a specific statistical standard. Specifically, NRC noted that the term "reasonable assurance" is not meant to imply a requirement that extreme values be used in analyses or that compliance be based on extreme values of predicted dose distributions. NRC also noted that the term "reasonable assurance" was not meant to indicate a significantly different standard than would be indicated by the term "reasonable expectation." Regardless of the terminology applied, the review's objective is to develop confidence that public health and safety will be protected.

This risk-informed and performance-based guidance document ensures that the NRC review is focused on those aspects most important to health and safety. The staff intends to incorporate risk insights into the review process and into the development of the staff's conclusions. Because this guidance document will be used to review a potentially large number of different types of waste determinations, its scope is general enough to allow the NRC staff to apply the guidance to a wide range of DOE waste determinations while also containing sufficient detail to ensure thoroughness and consistency of the staff's reviews.

Section 1 of this guidance document covers the site information needed to provide context to the reviewer and to support a performance assessment. Section 2 discusses the incidental waste criteria. Section 3 provides information on assessing removal of radionuclides and on estimating waste classification. Sections 4–7 address the performance objectives of

10 CFR Part 61, Subpart C. Section 4 of the document provides guidance for the staff review of the long-term performance assessment to evaluate the protection of the general population (10 CFR 61.41). Section 5 provides guidance for the staff review of assessments of potential radiation exposures to an inadvertent intruder (10 CFR 61.42). The review procedures presented in Section 6 are for assessing compliance with the radiation protection standards during operations (10 CFR 61.43). Section 7 of the document provides guidance for assessing long-term site stability (10 CFR 61.44). Guidance for assessing the implementation of DOE quality assurance programs is in Section 8, and a brief summary of how the NRC review will be conducted and documented is provided in Section 9. Information regarding monitoring activities is provided in Section 10.

Structure of the Staff Guidance Document

Where applicable, sections of this guidance document are divided into two subsections that describe the steps of the review process.

Areas of Review. This subsection describes the scope of the review (i.e., what is to be reviewed). It briefly discusses the specific types of technical information and analyses that should be reviewed. The areas listed are intended to be used for broad application; therefore, the listing is not exhaustive and may be supplemented for a given review, as appropriate.

Review Procedures. This subsection discusses the appropriate review topics and techniques. The reviewer should generally determine that the topics listed in this section are evaluated or that there is adequate technical basis to conclude that a specific topic does not need to be addressed. The reviewer should evaluate whether the information provided is sufficient to support the conclusions presented in the waste determination.

This guidance document covers a variety of site conditions and facility designs. Pertinent review procedures are contained in each section. Because the reviews are conducted on a case-by-case basis, the reviewer may emphasize particular aspects of each guidance document section as appropriate. Where possible, the proposed review procedures are based on applicable existing NRC guidance and previous NRC experience gained from DOE waste determination reviews.

Updating the Staff Guidance Document

This guidance document will be revised and updated periodically to clarify the content or incorporate modifications as the need arises. A revision number and publication date will be issued as needed.

Background

The concept of incidental waste, also known as waste-incidental-to-reprocessing (WIR) or non-high-level waste (non-HLW), is that some wastes can be managed based on their risk to human health and the environment, rather than based on the origin of the wastes. With respect to wastes resulting from the reprocessing of spent nuclear fuel, such as the tank residuals at some DOE sites, some are highly radioactive and need to be treated and disposed of as HLW in a geologic repository, but others do not. Incidental waste does not pose the same amount of risk to human health and the environment as HLW and does not need to be disposed of as HLW

1 to manage the risks that it poses. Consequently, incidental waste is not considered to be HLW.
2 DOE uses technical analyses documented in a "waste determination" to evaluate whether waste
3 is incidental or HLW. A waste determination is DOE's analysis as to whether the waste will
4 meet the applicable incidental waste criteria and usually includes a performance assessment. A
5 performance assessment is a quantitative evaluation of potential releases into the environment
6 and the resultant radiological doses, and the evaluation is often performed using a computer
7 model (or multiple models).
8
9 The concept of incidental waste has been recognized since 1969 when the Atomic Energy
10 Commission, NRC's predecessor agency, issued for comment a draft policy statement
11 regarding the siting of reprocessing facilities in the form of a proposed Appendix D to
12 10 CFR Part 50, which addressed a definition of HLW (AEC, 1969). The draft policy statement
13 provided that certain materials resulting from reprocessing could be disposed of in accordance
14 with 10 CFR Part 20 requirements. Although the draft policy statement did not use the term
15 "incidental," the Commission intended that the term HLW not include certain wastes that were
16 incidental to reprocessing operations. However, when Appendix D was finalized as Appendix F,
17 it did not include the paragraphs on incidental waste, because the Commission wanted to
18 preserve its flexibility as to how such material should be treated. The term "incidental waste"
19 was apparently first used in NRC's 1987 advance notice of proposed rulemaking to redefine the
20 definition of HLW (NRC, 1987). However, in the 1989 final rulemaking action on disposal of
21 radioactive waste, the Commission did not redefine HLW (NRC, 1989).
22
23 In 1990, the States of Oregon and Washington petitioned the Commission to amend
24 10 CFR Part 60 to redefine HLW. The petition concerned whether Hanford tank waste was
25 subject to NRC licensing jurisdiction. In response to the petition, the Commission approved
26 specific criteria for determining whether waste was incidental and issued a Staff Requirements
27 Memorandum (SRM) dated February 16, 1993, in response to SECY-92-391, "Denial of
28 PRM 60-4: Petition for Rulemaking from the States of Washington and Oregon Regarding
29 Classification of Radioactive Waste at Hanford." NRC published the criteria in the *Federal*
30 *Register* as part of the petition denial, as follows (NRC, 1993):
31
32   (1) The waste has been processed (or will be further processed) to remove
33   key radionuclides to the maximum extent that is technically and
34   economically practical,
35
36   (2) The waste will be incorporated in a solid physical form at a concentration
37   that does not exceed the applicable concentration limits for Class C
38   low-level radioactive waste (LLW) as set out in Title 10 of the *Code of*
39   *Federal Regulations* (10 CFR) Part 61, and
40
41   (3) The waste is to be managed, pursuant to the Atomic Energy Act, so that
42   safety requirements comparable to the performance objectives set out in
43   10 CFR Part 61, are satisfied.
44
45 The performance objectives of 10 CFR Part 61, Subpart C, include provisions for protecting the
46 general population from releases of radioactivity, protecting individuals from inadvertent
47 intrusion, and protecting individuals during operations. The performance objectives also include
48 provisions for the stability of the site after closure.
49

In a May 30, 2000, SRM on SECY-99-0284, "Classification of Savannah River Residual Tank Waste as Incidental," the Commission indicated that a more performance-based approach should be taken to determine whether waste could be classified as incidental (NRC, 2000b). In effect, cleanup to the maximum extent that is technically and economically practical and demonstration that performance objectives could be met (consistent with those which the Commission demands for the disposal of LLW) should adequately protect the public health and safety and the environment. In the Final Policy Statement for the Decommissioning Criteria for the West Valley Demonstration Project at the West Valley Site, the Commission adopted this performance-based approach and stated the criteria that should be applied to the incidental waste determinations at West Valley, as follows (NRC, 2002b):

    (1) The waste should be processed (or should be further processed) to remove key radionuclides to the maximum extent that is technically and economically practical; and

    (2) The waste should be managed so that safety requirements comparable to the performance objectives in 10 CFR Part 61, Subpart C, are satisfied.

In July 1999, DOE issued DOE Order 435.1, "Radioactive Waste Management" and the associated Manual, DOE M 435.1-1, "Radioactive Waste Management Manual," both of which were subsequently revised in 2001 (DOE 2001a,b). DOE M 435.1-1 states that waste determined to be incidental to reprocessing is not HLW and shall be managed in accordance with the requirements for TRU waste or LLW if it meets appropriate criteria. DOE M 435.1-1 discusses DOE's incidental waste evaluation process and the criteria for determining whether waste is incidental to reprocessing.

Prior to the passage of the NDAA (see the discussion that follows), DOE had periodically requested that NRC provide technical reviews of specific waste determinations. NRC provided technical assistance and advice to DOE regarding DOE's waste determinations and did not provide regulatory approval for DOE's actions. NRC's reviews were generally performed by the Office of Nuclear Material Safety and Safeguards under site-specific reimbursable Interagency Agreements (IAs) and Memoranda of Understanding (MOUs).

The staff reviewed DOE's waste determinations to assess whether DOE's technical assumptions, analyses, and conclusions were reasonable and whether there was reasonable assurance that the applicable criteria could be met. In general, the staff examined technical areas such as estimated radionuclide inventory, technology alternatives, performance assessment methodology, engineered system performance, infiltration, release and transport parameters, receptor scenarios and assumptions, and uncertainty and sensitivity analysis. The staff typically evaluated DOE-submitted information, generated requests for additional information (RAIs), met with DOE representatives to discuss technical questions and issues, and documented the final review results in Technical Evaluation Reports (TERs). Typically, the staff provided the associated MOUs, IAs, and TERs to the Commission for review before taking action. In addition to the review of the SRS tank closure methodology discussed above (NRC, 1999), the staff developed Commission papers for reviews of incidental waste determinations for waste intended to be removed from tanks at Hanford (NRC, 1997b), sodium-bearing wastes at the Idaho National Laboratory (INL) (NRC, 2002c), and tank farm closure at INL (NRC, 2003b). After completing any Commission-directed changes, the NRC staff transmitted the final TERs to DOE (NRC, 1997a, 2000a, 2002a, 2003a).

The Ronald W. Reagan National Defense Authorization Act for Fiscal Year 2005

In 2004, Senator Lindsey Graham of South Carolina introduced legislation that would allow DOE to use a process similar to the incidental waste process in DOE Order 435.1 at the Savannah River Site (SRS). During the development of the legislation, Congress inquired about the NRC's position on incidental waste and the proposed legislation, and the Commission responded in letters to Senator Inhofe on May 18, 2004, and Senator Jeffords on July 15, 2004 (NRC, 2004a,b).

Congress passed the NDAA on October 9, 2004, and the President signed it on October 28, 2004. Section 3116 of the NDAA allows DOE to continue to use an incidental waste process to determine that waste is not HLW; however, the NDAA is applicable to only South Carolina and Idaho and does not apply to waste transported out of these States. The NDAA requires that (1) DOE consult with NRC on its non-HLW determinations and plans and (2) NRC, in coordination with the State, monitor disposal actions DOE takes to assess compliance with NRC regulations in 10 CFR Part 61, Subpart C. If NRC determines that any disposal actions DOE takes are not in compliance, NRC shall inform DOE, the State, and Congressional subcommittees as soon as practicable. In addition, the NDAA provides for judicial review of any failure of NRC to carry out its monitoring responsibilities. The NDAA is provided in Appendix A of this guidance document.

The criteria contained in the NDAA for determining whether waste is non-HLW are similar to the incidental waste criteria previously used by NRC and specify that such waste

  (1) Does not require permanent isolation in a deep geologic repository for spent fuel or HLW;

  (2) Has had highly radioactive radionuclides removed to the maximum extent practical; and

  (3)A Does not exceed concentration limits for Class C low-level waste (LLW) and will be disposed of in compliance with the performance objectives in 10 CFR Part 61, Subpart C; or

  (3)B Exceeds concentration limits for Class C LLW but will be disposed of in compliance with the performance objectives of 10 CFR Part 61, Subpart C, and pursuant to plans developed by DOE in consultation with the NRC.

After NDAA enactment, the NRC staff developed an implementation plan that described how the staff would carry out its new responsibilities. That plan was described in SECY-05-0073, dated April 28, 2005 (NRC, 2005d). The Commission commented on and approved the staff's proposed plans in an SRM dated June 30, 2005 (NRC, 2005e).

Role of the U.S. Nuclear Regulatory Commission

The four DOE sites with potential incidental waste are operating under different requirements for waste evaluation and management. The NRC anticipates that SRS and INL will use the NDAA requirements for waste being disposed of in the State, but could decide to use DOE Order 435.1

1　for waste being shipped out of the State or possibly for certain wastes remaining in the State
2　that are not covered by the NDAA (e.g., waste not covered by a State-issued closure plan or
3　permit, as stated in the NDAA). Hanford is not covered under the NDAA and could decide to
4　use the requirements of DOE Order 435.1. West Valley will use NRC's Final Policy Statement
5　for the Decommissioning Criteria for the West Valley Demonstration Project (WVDP) at the
6　West Valley Site (the West Valley Policy Statement) for waste being disposed of onsite, and
7　could decide to use DOE Order 435.1 for waste being sent offsite. Alternatively, DOE may
8　decide to apply the NDAA requirements to all of its sites for consistency. The role of NRC and
9　the scope of the staff's reviews may vary depending on the criteria being applied in a specific
10　waste determination.
11
12　The NDAA applies only to waste remaining in South Carolina and Idaho. Under the NDAA, the
13　NRC has a statutory obligation to provide consultation to DOE for such waste. NRC performs a
14　technical review of DOE's waste determinations and arrives at an independent conclusion as to
15　whether there is reasonable assurance that the NDAA criteria can be met by DOE's waste
16　management approach. However, NRC does not have a regulatory role and the final decision
17　on whether incidental waste meets the criteria rests with the DOE. Guidance on performing
18　technical reviews under the NDAA is provided in this guidance document. Also, NRC must
19　monitor DOE's disposal actions for such waste to assess compliance with the performance
20　objectives of 10 CFR Part 61, Subpart C. The NRC must report any noncompliance to
21　Congressional subcommittees, the State, and DOE as soon as practicable after discovery of the
22　noncompliance. Monitoring is discussed in more detail in Section 10.
23
24　For the West Valley site, NRC's Final Policy Statement for the Decommissioning Criteria for the
25　WVDP at the West Valley Site applies to any residual material remaining at the site, including
26　any incidental waste (NRC, 2002b). Currently at West Valley, DOE is acting as a surrogate for
27　the licensee (the New York State Energy Research and Development Authority), and NRC is in
28　an advisory role with respect to DOE's incidental waste determinations. In providing the
29　incidental waste criteria, the West Valley Policy Statement states that "it is the Commission's
30　expectation that it will apply this criteria at the WVDP at the site following the completion of
31　DOE's site activities" (NRC, 2002b). Guidance on performing technical reviews under the West
32　Valley Policy Statement is provided in this guidance document. For offsite waste disposal, it is
33　DOE's responsibility to determine which criteria are applicable; for example, DOE may decide to
34　apply DOE Order 435.1. Guidance on technical reviews of waste determinations that DOE
35　decides to perform under DOE Order 435.1 also is provided in this document. For waste
36　determinations that DOE decides to perform using DOE Order 435.1 at the West Valley site,
37　NRC provides technical advice and consultation in an advisory manner.
38
39　At Hanford, DOE is responsible for determining which criteria are applicable to incidental waste
40　determinations. DOE, the U.S. Environmental Protection Agency, and the State of Washington
41　Department of Ecology entered into the Hanford Federal Facility Agreement and Consent Order
42　(the Tri-Party Agreement) in 1989. Appendix H of the Tri-Party Agreement requires that DOE
43　"establish an interface with the [NRC], and reach formal agreement on the retrieval and closure
44　actions for single shell tanks with respect to allowable waste residuals in the tank and soil
45　column" for those tanks for which DOE could not remove 99 percent of the waste by volume
46　(DOE, 1989). NRC provides technical advice and consultation in an advisory manner for waste
47　determination reviews performed for the Hanford site, as requested.
48

References

Citizens Advisory Board, Savannah River Site (CAB). "Comments on NRC Standard Review Plan Scope." Letter from J. Sulc and R. Meisenheimer to A. Bradford, NRC. November 2005.

Greeves, J. and J. Lieberman. "Comments on U.S. Nuclear Regulatory Commission Draft Interim Concentration Averaging Guidance." E-mail from J. Greeves and J. Lieberman to M. Lesar, NRC. April 2006.

South Carolina Department of Health and Environmental Control (SCDHEC). "Comments from the South Carolina Department of Health and Environmental Control on the Nuclear Regulatory Commission Scoping of the Standard Review Plan for Waste Determination Reviews." E-mail from S. Sherritt to A. Bradford, NRC. November 2005.

Idaho Department of Environmental Quality (Idaho). "State of Idaho Comments on the Draft Interim Concentration Averaging Guidance for Waste Determinations." E-mail from B. Olenick to A. Bradford, NRC. January 2006.

Natural Resources Defense Council (NRDC). "Comments of the Natural Resources Defense Council on the Nuclear Regulatory Commission's Draft Interim Concentration Averaging Guidance for Waste Determinations." E-mail from G. Fettus and M. McKinzie to A. Bradford, NRC. January 2006.

Oregon Department of Energy (Oregon). "Docket Number PROJ0734, PROJ0735, PROJ0736, and POOM-32." Letter from K. Niles to A. Bradford, NRC. January 2006.

U.S. Atomic Energy Commission (AEC). "Siting of Commercial Fuel Reprocessing Plants and Related Waste Management Facilities." *Federal Register*, 34 FR 8712. June 1969.

U.S. Department of Energy (DOE). "Hanford Federal Facility Agreement and Consent Order." May 1989.

———. DOE Order 435.1, "Radioactive Waste Management." DOE O 435. August 2001a.

———. DOE Order 435.1, "Radioactive Waste Management Manual." DOE M 435.1-1. June 2001b.

U.S. Nuclear Regulatory Commission (NRC). "Definition of High-Level Radioactive Waste, Advanced Notice of Proposed Rulemaking." *Federal Register*, 52 FR 5992. February 1987.

———. "Disposal of Radioactive Wastes, Final Rule." *Federal Register*, 54 FR 22578. May 1989.

———. "Reasonable Assurance: [Disposal of High Level Radioactive wastes in a Proposed Geologic repository at Yucca Mountain, Nevada; Final rule." *Federal Register*, 66 FR 55740. November, 2001.

———. "Denial of Petition for Rulemaking: States of Washington and Oregon." *Federal Register*, 58 FR 12342. March 1993.

1 ——. "Classification of Hanford Low-Activity Tank Waste Fraction as Incidental." Letter from
2 C. Paperiello to J. Kinzer, DOE. June 1997a.
3
4 ——. "Classification of Hanford Low-Activity Tank Waste Fraction as Incidental."
5 SECY–97–083. April 1997b.
6
7 ——. "Classification of Savannah River Residual Tank Waste as Incidental." SECY–99–284.
8 December 1999.
9
10 ——. "Savannah River Site High-Level Waste Tank Closure: Classification of Residual
11 Waste as Incidental." Letter from W. Kane to R.J. Schepens, DOE. June 2000a.
12
13 ——. "Staff Requirements—SECY–99–0284—Classification of Savannah River Residual
14 Tank Waste as Incidental." SRM-SECY-99-0284. May 2000b.
15
16 ——. "NRC Review of Idaho National Engineering and Environmental Laboratory Draft Waste
17 Incidental to Reprocessing Determination for Sodium-Bearing Waste - Conclusions and
18 Recommendations." Letter from J. Greeves to J. Case, DOE. August 2002a.
19
20 ——. "Final Policy Statement for the Decommissioning Criteria for the West Valley
21 Demonstration Project at the West Valley Site." Federal Register, 67 FR 5003. February
22 2002b.
23
24 ——. "NRC Review of Idaho National Engineering and Environmental Laboratory Draft
25 Incidental Waste (Waste Incidental to Reprocessing) Determination for Sodium-Bearing Waste."
26 SECY–02–0112. June 2002c.
27
28 ——. "NRC Review of Idaho Nuclear Technology and Engineering Center Draft Waste
29 Incidental to Reprocessing Determination for Tank Farm Facility Residuals—Conclusions and
30 Recommendations." Letter from L. Kokajko to J. Case, DOE. June 2003a.
31
32 ——. "NRC Review of Idaho National Engineering and Environmental Laboratory Draft
33 Incidental Waste (Waste-Incidental-to-Reprocessing) Determination for Tank Farm Facility
34 Closure." SECY–03–0079. May 2003b.
35
36 ——. Letter from N. Diaz, NRC Chairman, to Senator J. Inhofe. May 2004a.
37
38 ——. Letter from N. Diaz, NRC Chairman, to Senator Jeffords. July 2004b.
39
40 ——. "Notice of Public Scoping Meeting and Solicitation of Scoping Comments Related to the
41 Standard Review Plan for Waste Determination Reviews." Federal Register, 70 FR 66472.
42 November 2005a.
43
44 ——. "Waste Determination Standard Review Plan Public Meeting." November 2005b.
45
46 ——. "Draft Interim Concentration Averaging Guidance for Waste Determinations." Federal
47 Register, 70 FR 74846. December 2005c.
48

1   ——. "Implementation of New U.S. Nuclear Regulatory Commission Responsibilities Under
2   the National Defense Authorization Act of 2005 in Reviewing Waste Determinations for the
3   U.S. Department of Energy." SECY–05–0073. April 2005d.
4
5   ——. "Staff Requirements—SECY–05–0073—Implementation of New U.S. Nuclear
6   Regulatory Commission Responsibilities Under the National Defense Authorization Act of 2005
7   in Reviewing Waste Determinations for the USDOE." SRM–SECY–05–0073. June 2005e.
8
9   Washington State Department of Ecology (Washington). "Comments on the USNRC Standard
10 Review Plan." Email from S. Dahl to A. Bradford, NRC. November 2005.
11
12 ——. "Washington Department of Ecology Comments on Docket Numbers PROJ0734,
13 PROJ0735, PROJ0736, POOM-32." Letter from S. Dahl to A. Bradford, NRC. February 2006.
14
15 Washington Closure Hanford (WCH). "Draft Interim Concentration Averaging Guidance for
16 Waste Determinations." Letter from P. Pettiette to A. Bradford, NRC. February 2006.

1 SITE-SPECIFIC AND GENERAL INFORMATION

1.1   Site-Specific System Description Information

This section of the guidance document describes information about a site, disposal facilities,[1] and waste management activities that a reviewer should evaluate at the beginning of a waste determination review. A reviewer should establish the proper context for the detailed technical review that should be performed according to the guidance provided in Sections 2–8 of this document. To accomplish this objective, the reviewer must understand how the proposed waste management activities, disposal facility design, natural site characteristics, performance assessment analyses, and inadvertent intruder analyses are used to support decisions based on the waste determination. If practical, the reviewer should visit the site during the review. The site visit should provide the reviewer with an opportunity to see many of the site features and is expected to facilitate the review of information related to the waste determination.

1.1.1   Brief Description and Scope

First, the reviewer should assess the decisions to be made based on the waste determination and identify the technical bases for those decisions. Thus, the reviewer should evaluate the purpose and scope of the proposed waste management activities, the U.S. Department of Energy's (DOE's) understanding of the applicable waste criteria for the waste determination (see Section 2), and how the proposed waste management activities meet the applicable criteria. Based on previous waste determinations, NRC anticipates that DOE may provide this information in one or more summary documents (e.g., as in DOE, 2005a–c; Rosenberger, et al., 2005; Buice, et al., 2005). Specifically, before beginning a detailed technical review, the reviewer should be familiar with the following information:

- A brief description of the disposal site and surrounding areas, including the size of the disposal facility and the location of the disposal facility on the larger DOE site;

- DOE's definition of the scope of the waste determination, including an identification of the waste streams to which decisions based on the waste determination will apply;

- A description of the purpose of the proposed waste management activities;

- A description of the major structures, systems, and components of the proposed disposal facilities;

- The proposed schedule for relevant waste management activities;

---

[1]As described in 10 CFR 61.2, the term "land disposal facility" means the land, building, structures, and equipment that are intended to be used for the disposal of radioactive wastes. The term "disposal site" refers to the portion of the land disposal facility that is used for disposal of waste. In this guidance document, unless otherwise specified, the scope of the terms "site" and "facility" is limited to those locations and structures immediately related to the DOE waste determination under review, and the features that could affect the performance of the proposed disposal facility (e.g., aquifers, rivers). Unless otherwise noted, the term "site" is not intended to apply to the much larger area under DOE control and ownership, such as the Savannah River Site or Idaho National Laboratory, except to the extent that features of the larger DOE site may affect the performance of the proposed disposal facility. In this guidance, the term "DOE site" refers to the entire area under DOE control.

- Which criteria are applicable to the waste determination (see Section 2); and

- An assessment of compliance with the applicable waste criteria, including the following:

  — A list of the radionuclides considered to be highly radioactive radionuclides in the context of the waste determination and a summary of the basis for their selection (see Sections 2.4.2 and 3.2);

  — A summary of waste removal activities, including inventories of radionuclides before and after removal (see Section 3);

  — The waste classification assumed by DOE, if applicable (see Section 3.5);

  — A summary of results of the performance assessment and inadvertent intruder analyses used to demonstrate compliance with the applicable performance objectives (see Sections 4 and 5);

  — The major features of the proposed activities that will ensure the safety of individuals during operations (see Section 6);

  — A summary of the pathways and radionuclides that dominate predicted doses to members of the public and workers during operations and to members of the public (including inadvertent intruders) after site closure (see Sections 4, 5, and 6);

  — A summary of natural and engineered features of the disposal site that significantly limit or prevent potential doses to individuals after site closure (see Section 4.6.1.1); and

  — The major features of the natural system and relevant waste disposal facilities that could either affect or ensure disposal site stability (see Section 7).

1.1.2   Facility Description

The reviewer should evaluate the facility description to understand the role of the disposal facility in the performance assessment, inadvertent intruder analysis, demonstration of the protection of individuals during operations, and demonstration of disposal site stability. In general, the areas of review are expected to encompass, but not necessarily be limited to, the technical information described in 10 CFR 61.12(b), (c), (d), (g), and (i). Specifically, the reviewer should evaluate the following information:

- A scale drawing or map of the relevant disposal unit(s) showing locations of radioactivity within the disposal unit(s), including a description of structures (e.g., equipment, tanks, pumps, piping, or disposal vaults) at the facility that are the subject of the waste determination;

- A system description, including the geometry of structures (e.g., tanks, contaminated equipment, or waste treatment facilities) and barriers (e.g., caps or vaults) that may be important in developing conceptual models to assess the performance of the facility;

- A description of the major design features of the disposal facility and disposal units and the relationship between the design features and performance objectives, including the following:

  — A description of design features related to the infiltration of water, integrity of covers for disposal units, and disposal site drainage;

  — A description of the design features related to disposal site closure including features designed to limit the need for long-term maintenance and to limit the potential for inadvertent intrusion;

  — A description of design features related to disposal site stabilization, including the structural stability of backfill, wastes, and covers;

- A description of the relationship between natural events and processes (see Section 1.1.3) and the principal facility design criteria;

- A description of the physical and chemical forms of the radionuclides;

- Information about past waste management activities (e.g., addition of chelating agents, waste treatment processes) to the extent that they may affect contaminant fate and transport modeling or affect monitoring activities; and

- Information about previous waste releases that is relevant to potential release pathways or to contamination that may affect proposed waste management activities.

In some instances, existing facilities, and especially older facilities, may not be adequately described. If the information provided is incomplete, the reviewer should determine what conservative assumptions are necessary to assess long-term facility performance.

1.1.3   Site Description

The reviewer should evaluate the natural and anthropogenic characteristics of the proposed disposal site that may affect its ability to meet the performance objectives of 10 CFR Part 61, Subpart C. It is expected that the reviewer will need the information described in this section to evaluate conceptual models supporting a performance assessment (e.g., hydrologic characteristics influencing radionuclide transport), inadvertent intruder analysis (e.g., the presence of natural resources that could influence methods of inadvertent intrusion), evaluation of doses to individuals during operations (e.g., the location of members of the public during operations), and evaluation of site stability (e.g., the nature of meteorological, hydrologic, and seismic disruptive events). It is anticipated that (1) the effect of individual natural and anthropogenic features on waste isolation will be different at different disposal sites and (2) the level of description provided may vary accordingly. In general, the areas of review are expected to encompass, but not necessarily be limited to, the technical information described in 10 CFR 61.12(a), (d), and (h). Additional guidance on reviewing these types of information is provided in other documents (see NRC, 2003a, 2000, 1988, 1982, 1981).

1.1.3.1 Site Location and Description

The reviewer should evaluate information related to the site location to understand the relationship between the disposal site and regional features and to establish potential receptor locations during operations and after site closure. To the extent possible, the reviewer should verify the information with independent reports [e.g., U.S. Geological Survey (USGS) maps, site maps, previous waste determinations, site planning documents]. Specifically, the reviewer should evaluate the following information:

- A scale drawing or map of the waste disposal system and related existing or proposed structures, showing sizes and locations on the DOE site;

- A scale drawing or map showing the location of the disposal site relative to prominent natural features such as rivers and lakes;

- A map that shows the topography (including elevations) of the site;

- A description of the manmade features of the site that may affect the waste isolation characteristics of the disposal site, such as parking lots or roads that may affect runoff and recharge;

- A description of neighboring property surrounding the DOE site;

- A description of biological characteristics of the area that may influence waste isolation (e.g., a description of any plants native to the region that could compromise closure caps through root intrusion or the presence of burrowing animals that could compromise closure caps or exhume waste); and

- A scale drawing or map that shows the location of members of the public before and after site closure and before and after the period of institutional controls ends, including the location of inadvertent intruders after the period of institutional controls ends.

1.1.3.2 Land Use

The reviewer should evaluate past, current, and potential future land use around the site to provide context for assumptions about the activities of members of the public (including potential inadvertent intruders) after site closure and activities of members of the public during operations. The reviewer should verify that DOE has used appropriate available data on land use to support its performance assessment and inadvertent intruder analyses. Specifically, the reviewer should perform the following review procedures:

- The reviewer should evaluate land uses prior to Federal control of the site and determine whether any changes to the site have been made that would preclude the resumption of similar activities (e.g., depletion of an aquifer).

- The reviewer should assess the current uses of land neighboring the DOE site and the impact these uses may have on future use of the disposal site after the end of institutional controls (e.g., the growth of a major metropolitan area neighboring the DOE site could limit agricultural uses of the disposal site).

- The reviewer should assess the current uses of land neighboring the DOE site and any impact the current land uses may have on assumptions about the location of members of the public during operations.

- The reviewer should evaluate a description of ground and surface waters on or near the site including information on resource type, occurrence, location, current projected uses, and potential future uses (e.g., agricultural, industrial, drinking water source).

- The reviewer should evaluate a description of the natural resources occurring at or near the site (e.g., metallic and nonmetallic ores, fossil fuels and hydrocarbons, and mineral deposits) and assess whether the exploitation of natural resources at the site could affect future development of land surrounding the disposal site.

- The reviewer should evaluate a description of these natural resources occurring at or near the site and assess whether the exploitation of the resources could affect the likelihood or potential methods of inadvertent intrusion into the disposal site.

1.1.3.3 Meteorology and Climatology

The reviewer should evaluate a description of the meteorology and climatology in the vicinity of the site to determine how local weather patterns could affect the site's ability to meet the performance objectives in 10 CFR Part 61, Subpart C. Thus, the reviewer should examine meteorological information necessary to support estimates of infiltration, airborne dose pathway analyses, and potential disruption of engineered barriers by processes such as flooding, erosion, and frost heaving. Because of the long timeframes to be considered, the reviewer also should verify whether DOE has provided a description of its projections of future climate changes and appropriately incorporated these projections into the performance assessment. Specifically, the reviewer should verify that DOE provided the following information:

- A description of the general climate of the region;

- A description of aspects of the local (site) meteorology that may affect performance assessment and dose calculations (e.g., temperature, precipitation intensity and duration, wind speed, wind direction, and atmospheric stability);

- A description of the projected future climate states to be considered in the performance assessments, including the basis for and uncertainties in those projections; and

- A description of the sensitivity of system performance to uncertainty in future climate states.

To the extent possible, the reviewer should verify site-specific information about meteorology and climatology against independent information (e.g., information provided by the National Weather Service or the National Oceanic and Atmospheric Administration). Projected future climate states should be supported by an adequate technical basis and compared to independent assessments (e.g., paleoclimate modeling performed by the National Climatic Data Center) to the extent possible. Because capabilities to predict climate change continue to evolve, additional information about climate change may be relevant if system performance is sensitive to uncertainties in future climate states.

1.1.3.4   Geology and Seismology

Information about geologic processes (e.g., earthquakes, erosion, faulting) that occur at the site is needed to support an assessment of site stability and to support the development of conceptual models used in the performance assessment and inadvertent intruder analysis. To support an assessment of how the geological and geotechnical characteristics of a site will affect its performance, the reviewer should evaluate the following information:

*   A description of the surface and subsurface geologic characteristics and stratigraphy of the site and its vicinity;

*   A description of the geomorphology of the site, including USGS topographic maps that emphasize local geomorphic features pertinent to the site, particularly for processes such as erosion that may affect long-term site stability;

*   A description of past seismic and volcanic activity at and near the site, the features and processes that caused the activity, and the potential for significant seismic and volcanic activity to occur in the future;

*   A description of the potential effects of seismic activity on site performance, including the effects of landslides, land movement, and liquefaction;

*   A description of manmade geologic features, such as mines or quarries, that may affect water runoff and recharge; and

*   A description of the structural stability of geotechnical features of the disposal facility (e.g., slope stability, potential for subsidence of backfilled soils that could affect cap stability).

To the extent possible, the reviewer should verify descriptions of the site-specific geological and seismological information with independent information, such as maps and other geospatial USGS-generated data.

1.1.3.5   Hydrology

The reviewer should use information about the surface water and groundwater hydrology to evaluate conceptual models of how the hydrologic characteristics at the site will affect site stability and radionuclide release and transport. Depending on the specific site, the reviewer may need to analyze the potential impact of atypical hydrological conditions (e.g., floods). Specifically, the reviewer should evaluate the following information:

*   A description of natural drainage and surrounding watershed fluvial features and anthropogenic features that may influence surface hydrology and the potential for flooding at the site;

*   Water resource data, including maps, hydrographs, and stream records from other agencies (e.g., USGS and U.S. Army Corps of Engineers);

- A description of the surface water bodies at the site and surrounding areas, including the location, size, shape, and other hydrologic characteristics of streams, lakes, or coastal areas;

- A description of the saturated zone, including potentially affected aquifers, the lateral extent, thickness, water-transmitting properties, recharge and discharge zones, groundwater flow directions and velocities, and other information that can be used to support the conceptual model of the saturated zone;

- Physical parameters, such as storage coefficients, transmissivities, hydraulic conductivities, porosities, and intrinsic permeabilities;

- A description of the unsaturated zone, including descriptions of the lateral extent and thickness of permeable and impermeable zones; the presence, lateral extent, and thickness of perched water zones; potential conduits of anomalously high flux; and direction and velocity of unsaturated flow;

- Physical parameters, such as porosity, water content (including temporal variation); saturated hydraulic conductivity; characteristic relationships between water content, pressure head, and hydraulic conductivity; and hysteretic behavior during wetting and drying cycles; and

- The distribution coefficients (K_d) of the radionuclides of interest and the associated technical basis for their selection (e.g., site-specific values, literature compilations, geochemical models).

1.1.3.6   Radiological Status

Because the DOE sites relevant to this guidance document (Savannah River Site, Idaho National Laboratory, Hanford, and West Valley) have operated for decades, areas near the proposed disposal facility may already be contaminated with radioactive material. In general, contamination resulting from spills or other releases of radioactive material at the site is addressed through alternate regulatory processes and is not within the scope of waste determinations. However, existing radioactive contamination may affect or provide useful information about the waste management activities proposed in a waste determination. Specifically, the reviewer should evaluate information about groundwater contamination that may provide relevant information about radionuclide release pathways, may constrain contaminant fate and transport models, or that may complicate subsequent monitoring activities. Therefore, a reviewer should evaluate information about the current radiological status of the area near the proposed disposal facility and its environs, including the following information:

- A summary of areas near the proposed disposal facilities where releases of radioactive material occurred in the past;

- A description of the types, forms, activities, and concentrations of radionuclides involved in the release; and

1 • A scale drawing or map of the site, facilities, and environs showing the locations of
2 relevant previous releases, including features such as abandoned boreholes or
3 disturbed soil that may affect contaminant fate and transport.
4
5 To ensure information describing the area near the proposed disposal facility is complete, the
6 reviewer should evaluate the purpose, major attributes, summary conclusions, and regulatory
7 program under which studies of existing contamination near the disposal facility were evaluated.
8 If significant groundwater contamination exists, the reviewer should evaluate historical
9 information about plume movement for comparison with the results of site-specific groundwater
10 models used to support the performance assessment (see Section 4) and DOE's plans to
11 monitor and model plume movement for coordination with NRC monitoring activities (see
12 Section 10).
13

14   1.2   Applicable Sites and Waste Criteria

15
16 This guidance document has been developed to address the reviews of waste determinations
17 for the four DOE sites that may have incidental waste:
18
19 • Savannah River Site (SRS), Aiken, South Carolina;
20
21 • Idaho National Laboratory (INL), Idaho Falls, Idaho;
22
23 • Hanford Site (Hanford), Richland, Washington; and
24
25 • West Valley Demonstration Project (West Valley), West Valley, New York.
26
27 The reviewer must determine whether DOE is using the applicable waste criteria for a given
28 waste determination. The different sets of waste criteria are detailed in Section 2 of this
29 guidance document. Guidance for the detailed technical evaluation of whether there is
30 reasonable assurance that the applicable waste criteria can be met is provided in Sections 2–8
31 of this guidance document.
32
33 Based on previous waste determinations, it is anticipated that DOE will provide information that
34 addresses each of the waste criteria it considers to be applicable to a specific waste
35 determination. The staff should review the DOE waste determination to verify that the
36 applicable waste criteria have been addressed. For example, the staff should confirm whether
37 DOE has identified the following criteria for specific sites:
38
39 • The Ronald W. Reagan National Defense Authorization Act for Fiscal Year 2005 (NDAA)
40 (Public Law 108-375, 2004) is applicable to certain waste disposed of in South Carolina
41 or Idaho subject to other requirements of the NDAA (see Section 2.1);
42
43 • The waste incidental to reprocessing criteria from DOE Order 435.1 and its associated
44 Manual (DOE, 2001a,b) may apply to waste in South Carolina or Idaho that is not
45 covered by the NDAA, waste at Hanford, or waste being shipped offsite from the West
46 Valley Demonstration Project (see Section 2.2); and
47
48 • The waste incidental to reprocessing criteria identified in the Decommissioning Criteria
49 for the West Valley Demonstration Project at the West Valley Site; Final Policy

Statement (NRC, 2002) are applicable to waste being disposed of on site at the West Valley site (see Section 2.3).

1.3   Prior Waste Determinations

The reviewer should consider prior waste determination reviews for relevance to the technical evaluation at hand and determine how the reviews might support an assessment of the DOE demonstration of compliance with appropriate waste criteria. For example, the reviewer could examine NRC's Technical Evaluation Reports (TERs) for tank closure at INL (NRC, 2003b) and for salt waste disposal at SRS (NRC, 2005). As additional waste determination reviews are completed, additional documents can be considered for each of the four DOE sites previously noted.

To the extent practical, the staff should evaluate prior TERs both to gain insights from previous NRC conclusions and recommendations and to understand any major changes from prior analyses. Staff should particularly focus on the extent to which the prior reviews provide insights into the current DOE waste determination with regard to information such as, but not limited to, the following:

- Site characterization;

- Waste inventory and source term;

- Locations, descriptions, operating history, and radiological status of any facilities associated with the proposed waste management activities;

- Existing and proposed waste management strategies and methodologies;

- Existing and proposed performance assessment and dose modeling methodologies and results;

- Existing and proposed monitoring systems; and

- Related structures and processes DOE employed in previous waste management strategies.

This evaluation may include prior waste determination reviews identified by either DOE or the reviewer. To the extent practical, the reviewer should consider prior waste determinations from both the site being currently evaluated and other DOE sites, if relevant.

1.4   References

Buice, J.M., R.K. Cauthen, R.R. Haddock, B.A., Martin, J.A. McNeil, J.L. Newman, and K.H. Rosenberger. "Performance Objective Demonstration Document (PODD) for the Closure of Tank 19 and Tank 18 Savannah River Site." CBU–PIT–2005–00106. Rev. 1. Westinghouse Savannah River Company. 2005.

1   Public Law 108-375, "Ronald W. Reagan National Defense Authorization Act for Fiscal Year
2   2005." October 2004.
3
4   U.S. Department of Energy (DOE). "Radioactive Waste Management." DOE O 435.1.
5   August 2001a.
6
7   ———. "Radioactive Waste Management Manual." DOE M 435.1-1. June 2001b.
8
9   ———. "Draft Section 3116 Determination Salt Waste Disposal Savannah River Site."
10   DOE–WD–2005–001. DOE-Savannah River. March 2005a.
11
12   ———. "Draft Section 3116 Determination Idaho Nuclear Technology and Engineering Center
13   Tank Farm Facility." DOE/NE–ID–11226. DOE, Idaho Operations Office. 2005b.
14
15   ———. "Draft Section 3116 Determination for Closure of Tank 19 and Tank 18 at the Savannah
16   River Site." DOE–WD–2005–002. DOE-Savannah River. September 2005c.
17
18   U.S. Nuclear Regulatory Commission (NRC). "Draft Environmental Impact Statement on
19   10 CFR Part 61 Licensing Requirements for Land Disposal of Radioactive Waste."
20   NUREG–0782. Washington, DC. September 1981.
21
22   ———. "Site Suitability, Selection, and Characterization, Branch Technical Position—Low-Level
23   Waste Licensing Branch." NUREG–0902. April 1982.
24
25   ———. "Standard Review Plan for the Review of a License Application for a Low-Level
26   Radioactive Waste Disposal Facility." NUREG–1200. January 1988.
27
28   ———. "A Performance Assessment Methodology for Low-Level Radioactive Waste Disposal
29   Facilities: Recommendations of NRC's Performance Assessment Working Group."
30   NUREG–1573. October 2000.
31
32   ———. "Decommissioning Criteria for the West Valley Demonstration Project (M–32) at the
33   West Valley Site; Final Policy Statement." *Federal Register*. Vol. 67, No. 22. February 2002.
34
35   ———. "Consolidated NMSS Decommissioning Guidance." Vols. 1–3. NUREG–1757.
36   September 2003a.
37
38   ———. "NRC Review of Idaho Nuclear Technology and Engineering Center Draft Waste
39   Incidental to Reprocessing Determination for Tank Farm Facility Residuals—Conclusions and
40   Recommendations." Letter from L. Kokajko to J. Case, DOE. June 2003b.
41
42   ———. "Technical Evaluation Report for the U.S. Department of Energy, Savannah River Site
43   Draft Section 3116 Waste Determination for Salt Waste Disposal." Letter from L. Camper to
44   C. Anderson, DOE. December 2005.
45
46   Rosenberger, K.H., B.C. Rogers, and R.K. Cauthen. "Saltstone Performance Objective
47   Demonstration Document (U)." CBU–PIT–2005–00146. Rev. 0. Westinghouse Savannah
48   River Company. 2005.

2 INCIDENTAL WASTE CRITERIA

This section of the guidance document discusses the criteria against which a waste determination will be reviewed. This will enable U.S. Nuclear Regulatory Commission (NRC) reviewers to understand the criteria and determine how to evaluate the approaches that the U.S. Department of Energy (DOE) might use to show that the criteria can be met. This section provides information concerning the waste criteria from Section 3116 of the Ronald W. Reagan National Defense Authorization Act for Fiscal Year 2005 (NDAA) (Section 2.1), the DOE Order 435.1 "Radioactive Waste Management" and the associated manual "Radioactive Waste Management Manual" (Section 2.2), and the West Valley Policy Statement (Section 2.3). In general, there are several similarities between the different sets of criteria (e.g., all of the sets of criteria refer to the performance objectives of 10 CFR Part 61, Subpart C) and a few important differences (e.g., not all of the sets of criteria require that certain concentration limits be met). Where appropriate, consistent guidance for reviewing similar criteria is provided. Differences in criteria are discussed in Section 2.4. Sections 2.5–2.7 provide information on reviewing certain criteria, some of which is described in greater detail in Sections 3–7.

2.1 Criteria from the Ronald W. Reagan National Defense Authorization Act for Fiscal Year 2005 (Section 3116)

Section 3116 of the NDAA (Public Law 108-375, 2004) (see Appendix A) identifies the following criteria for determining that waste is not high-level waste (HLW):

(1) The waste does not require permanent isolation in a deep geologic repository for spent nuclear fuel or HLW (see Section 2.5);

(2) The waste has had highly radioactive radionuclides removed to the maximum extent practical (see Sections 2.6 and 3); and

(3)A The waste does not exceed concentration limits for Class C low-level waste (LLW) and will be disposed of in compliance with the performance objectives in 10 CFR Part 61, Subpart C (see Sections 2.7 and 3–7); or

(3)B The waste exceeds concentration limits for Class C LLW but will be disposed of in compliance with the performance objectives in 10 CFR Part 61, Subpart C (see Section 4.7), and pursuant to plans DOE developed in consultation with NRC (see Sections 2.7 and 3.7).

As described in paragraphs (c) and (d) of Section 3116 of the NDAA, these criteria apply to certain waste that will be disposed of in South Carolina and Idaho and not to waste that will be transported out of those States. The NDAA may not apply to waste that does not meet other criteria in the NDAA (e.g., waste not covered by State-issued closure plans or permits).

2.2 Criteria from DOE Order 435.1

DOE Order 435.1 and the related Manual 435.1-1 (DOE, 2001a,b) consider that incidental waste may be either LLW or transuranic (TRU) waste and include the following two corresponding sets of similar criteria for determining that waste is incidental to reprocessing. Incidental wastes that will be managed as LLW will meet the following criteria:

(1)   Have been processed or will be processed to remove key radionuclides to the maximum extent that is technically and economically practical (see Sections 2.6 and 3);

(2)   Will be managed to meet safety requirements comparable to the performance objectives set out in 10 CFR Part 61, Subpart C (see Sections 2.7 and 4.7); and

(3)   Are to be managed, pursuant to DOE's authority under the Atomic Energy Act of 1954, as amended, and in accordance with the provisions of Chapter IV of the DOE Radioactive Waste Management Manual (DOE, 2001b), provided the waste will be incorporated in a solid physical form at a concentration that does not exceed the applicable concentration limits for Class C LLW as set out in 10 CFR 61.55 (see Section 3.5) or will meet alternative requirements for waste classification and characterization as DOE may authorize.

Incidental wastes that will be managed as transuranic wastes will meet the following criteria:

(1)   Have been processed or will be processed to remove key radionuclides to the maximum extent that is technically and economically practical (see Sections 2.6 and 3);

(2)   Will be incorporated in a solid physical form and meet alternative requirements for waste classification and characteristics, as DOE may authorize; and

(3)   Are managed pursuant to DOE's authority under the Atomic Energy Act of 1954, as amended, in accordance with the provisions of Chapter III of the DOE Radioactive Waste Management Manual (DOE, 2001b), as appropriate.

DOE may decide to apply the requirements in DOE Order 435.1 and the associated Manual (DOE, 2001a,b) to waste in South Carolina or Idaho that is not covered by the NDAA, waste at Hanford, or waste being shipped offsite from West Valley.

The wording in DOE M 435.1-1 does not require that DOE necessarily use the waste classifications in 10 CFR 61.55 or the performance objectives in 10 CFR Part 61, and DOE has the flexibility to use criteria that are different (DOE, 2001b). NRC does not have a regulatory or statutory role with regard to whether the alternate criteria DOE proposed in a waste determination in accordance with DOE 435.1 are acceptable. Because the alternate criteria that DOE may propose cannot be known at this time, specific associated guidance cannot be provided in this guidance document; however, the general approach and review areas identified in the guidance document will still provide useful insights to the reviewer. In general, the NRC staff review should assess whether there is reasonable assurance that the proposed alternate criteria can be met and may provide an independent risk-informed, performance-based assessment of whether the criteria can protect public health and safety.

In February 2002, the Natural Resources Defense Council (NRDC), Snake River Alliance, and the Yakama and Shoshone-Bannock Nations filed suit against the DOE, stating that the Nuclear

1 Waste Policy Act did not allow DOE to reclassify HLW and dispose of it anywhere except in a
2 geologic repository. In July 2003, the Federal District Court in Idaho granted summary
3 judgment to NRDC and declared DOE's incidental waste process, as described in DOE
4 Order 435.1, invalid. DOE appealed the decision; in November 2004, the U.S. Court of Appeals
5 for the Ninth Circuit vacated the lower court's decision on ripeness grounds.
6
7 2.3 Criteria from the West Valley Policy Statement
8
9 The NRC's "Decommissioning Criteria for the West Valley Demonstration Project at the West
10 Valley Site; Final Policy Statement" (West Valley Policy Statement) identifies the following
11 criteria for incidental waste determinations (NRC, 2002):
12
13 (1) The waste should be processed (or should be further processed) to remove
14 key radionuclides to the maximum extent that is technically and economically
15 practical (see Sections 2.6 and 3); and
16
17 (2) The waste should be managed so that safety requirements comparable to the
18 performance objectives in 10 CFR Part 61, Subpart C, are satisfied (see
19 Sections 2.7 and 4.7).
20
21 NRC's West Valley Policy Statement provides criteria that are applicable to waste that may be
22 disposed of onsite at West Valley.
23
24 The West Valley Policy Statement also states that "the resulting calculated dose from incidental
25 waste is to be integrated with all the other calculated doses from the material remaining at the
26 entire NRC-licensed site." However, this guidance document covers only the review of
27 incidental waste determinations, not the larger analysis of whether the entire site meets the
28 West Valley Policy Statement or any other applicable requirements.
29
30 2.4 Comparison of Criteria
31
32 In general, there are several similarities between the different sets of criteria (e.g., all of the sets
33 of criteria refer to the performance objectives of 10 CFR Part 61, Subpart C) and a few
34 important differences (e.g., not all of the sets of criteria require that certain concentration limits
35 be met). Where appropriate, this guidance document provides consistent guidance for
36 reviewing those criteria that are similar, as discussed below.
37
38 2.4.1 "Waste Incidental to Reprocessing" and "Non-High-Level Waste"
39
40 Historically, the type of waste addressed in waste determinations has been referred to as
41 "waste-incidental-to-reprocessing" (WIR) or "incidental waste," and the waste determinations
42 have been called "WIR determinations." The NDAA does not use the term "incidental waste" or
43 "WIR" but instead specifies that HLW does not include wastes that meet the criteria of the
44 NDAA; therefore, DOE refers to waste that is covered by the NDAA as "non-HLW" and the
45 associated waste determinations as "non-HLW determinations." The NRC staff considers WIR,
46 incidental waste, and non-HLW to be the same type of waste; that is, they are wastes that are
47 incidental to the reprocessing of nuclear fuel and can be managed as LLW if the appropriate
48 criteria can be met. This guidance document uses the term "incidental waste" to mean both

1 "WIR" and "non-HLW waste" and uses the term "waste determinations" to mean both "WIR
2 determinations" and "non-HLW determinations."
3

2.4.2 "Highly Radioactive Radionuclides" and "Key Radionuclides"

5
6 The NDAA refers to "highly radioactive radionuclides," while DOE M 435.1-1 and the West
7 Valley Policy Statement both refer to "key radionuclides." The NRC staff has previously stated
8 that it believes that "highly radioactive radionuclides" are those radionuclides that contribute
9 most significantly to risk to the public, workers, and the environment (NRC, 2005). This is the
10 same concept as key radionuclides, as used in previous NRC reviews of waste determinations
11 (NRC, 2003). Therefore, for purposes of evaluating waste determinations, the NRC staff
12 considers key radionuclides and highly radioactive radionuclides to be equivalent. For ease of
13 reference, this guidance document uses the term "highly radioactive radionuclides" to also
14 mean "key radionuclides." See Section 3.2 for guidance on evaluating the identification of
15 highly radioactive radionuclides.
16

2.4.3 "Maximum Extent Practical" and "Maximum Extent Technically and Economically Practical"

19
20 The NDAA refers to removal of radionuclides to the "maximum extent practical," while
21 DOE M 435.1-1 and the West Valley Policy Statement both refer to "the maximum extent
22 technically and economically practical." The "maximum extent practical" is similar to the
23 "maximum extent technically and economically practical," but allows for somewhat broader
24 considerations of what is practical (e.g., DOE's schedule, programmatic considerations, other
25 risk considerations such as non-radiological risk). However, in most cases, those broader
26 considerations should be evaluated in a quantitative manner; for example, schedule delays
27 could be quantified by estimating the monetary cost or the risk of delaying waste processing.
28 NRC staff believes that DOE should consider both technological and economic aspects of
29 waste removal in demonstrating compliance with the maximum extent practical criterion. For
30 ease of reference, this guidance document will use the term "maximum extent practical" to also
31 mean "maximum extent technically and economically practical." See Section 3 for guidance on
32 evaluating the removal to the maximum extent practical.
33
34 NRC staff believes that the purpose of the various criteria related to radionuclide removal is to
35 minimize the inventory of highly radioactive radionuclides disposed of as incidental waste. In
36 many cases, the intent of requiring removal of highly radioactive radionuclides to the maximum
37 extent practical can be satisfied by reducing the volume of residual waste in a contaminated
38 structure (e.g., a tank, an evaporator) to the maximum extent practical. However, evaluating
39 alternative methods of physically removing waste from a structure does not eliminate the need
40 to consider (1) whether it would be practical to remove selected highly radioactive radionuclides
41 from the waste (e.g., by chemical extraction) or (2) whether it would be practical to remove the
42 contaminated structure for disposal instead of stabilizing it and disposing of it in place. In this
43 guidance, "removal" of radionuclides refers to removal of waste from a disposal system (either
44 by removing waste alone or by removing contaminated structures that contain waste) or
45 removal of radionuclides from a waste stream (e.g. treatment of salt waste at SRS to remove
46 highly radioactive radionuclides prior to disposing of the waste), as applicable.
47

2.4.4   Differences in Concentration Limits

The NDAA specifies that if the waste being evaluated exceeds the concentration limits for Class C waste, as given in 10 CFR 61.55, DOE must consult with NRC on the development of its disposal plans for that waste. Although additional consultation is required, the NDAA does not prohibit waste that exceeds Class C concentration limits from being determined to be incidental waste. Although the NDAA does not specify that the waste must be in solid form, as DOE M 435.1-1 does, NRC generally requires that waste be disposed of in solid form to provide stability (10 CFR 61.56), and the staff believes this is appropriate for waste that is evaluated in compliance with 10 CFR Part 61 stability requirements or comparable requirements.

DOE M 435.1-1 specifies that the waste must be in solid form at a concentration that does not exceed the applicable concentration limits for Class C LLW as set out in 10 CFR 61.55 or that the waste will meet alternative requirements for waste classification and characterization as DOE may authorize. Therefore, DOE M 435.1-1 does prohibit waste that exceeds Class C concentration limits from being determined to be incidental waste unless DOE authorizes alternate criteria. See Section 3.5 for guidance on reviewing the evaluation of the class of the waste. Because any alternate criteria that DOE may propose cannot be known at this time, specific associated guidance cannot be provided in this guidance document; however, the general approach and review areas identified in the guidance will still provide useful insights to the reviewer. If DOE does authorize alternate criteria and NRC is reviewing the associated waste determination, the reviewer should evaluate whether there is reasonable assurance that the alternate criteria can be met and whether the proposed alternate criteria are protective of public health and safety.

Although the West Valley Policy Statement does not include a concentration limit criterion with respect to determining whether waste is incidental, it does provide that the waste should be managed so that safety requirements comparable to the performance objectives in 10 CFR Part 61, Subpart C, are satisfied.

2.4.5   Alternatives to the Performance Objectives of 10 CFR Part 61, Subpart C

The NDAA specifies that the "waste ... will be disposed of in compliance with the performance objectives in 10 CFR Part 61, Subpart C." DOE M 435.1-1 and the West Valley Policy Statement require that the "waste should be managed so that safety requirements comparable to the performance objectives in 10 CFR Part 61, Subpart C, are satisfied." In other words, the NDAA does not allow for safety requirements that are "comparable" to 10 CFR Part 61, Subpart C, but instead specifies that only Subpart C can be used. For waste determinations made using DOE Order 435.1 and the West Valley Policy Statement, DOE could propose alternate safety requirements, as long as adequate technical basis is provided for determining that the proposed alternate safety requirements are comparable to 10 CFR Part 61, Subpart C. In general, because the requirements in Subpart C are NRC's performance objectives for the disposal of waste in LLW facilities, strong justification for using alternative safety requirements would be needed to demonstrate that public health and safety were protected, unless it can be determined that the proposed alternate safety requirements are more stringent than those of 10 CFR Part 61, Subpart C. Because the proposed alternative safety requirements that DOE may propose in the future cannot be known at this time, specific associated guidance cannot be provided in this guidance document; however, the general approach and review areas identified

1 in the guidance will provide useful insights to the reviewer. See Sections 4–7 for guidance on
2 reviewing whether there is reasonable assurance that the performance objectives of
3 10 CFR Part 61, Subpart C, can be met.
4

2.5   Does Not Require Disposal in a Geologic Repository

7 The NDAA contains a criterion that the waste does not require permanent isolation in a deep
8 geologic repository for spent fuel or high-level radioactive waste. DOE M 435.1-1 and the West
9 Valley Policy Statement do not contain the same criterion with respect to incidental waste
10 determinations, although other statutes or regulations may require HLW disposal in a
11 geologic repository.
12

2.5.1   Areas of Review

15 In general, there is reasonable assurance that this criterion can be met if the two other NDAA
16 criteria can be met. In other words, if highly radioactive radionuclides have been removed to the
17 maximum extent practical and the waste will be disposed of in compliance with the performance
18 objectives in 10 CFR Part 61, Subpart C (which are the same performance objectives NRC uses
19 for disposal of low-level waste), then this supports a conclusion that the waste does not require
20 disposal in a deep geologic repository. However, this criterion allows for the consideration that
21 waste may require disposal in a geologic repository even though the two other NDAA criteria
22 may be met. Those circumstances under which geologic disposal is warranted to protect public
23 health and safety and the environment could be considered; for example, unique radiological
24 characteristics of waste or nonproliferation concerns for particular types of material
25 (NRC, 2005).
26

2.5.2   Review Procedures

29 The reviewer should determine whether the following conditions are met:
30
31 •   There is reasonable assurance that the other criteria of the NDAA can be met (see
32   Sections 2–8); and
33
34 •   No other characteristics of the waste would necessitate that the waste be disposed of in
35   a deep geologic repository in order to protect public health and safety.
36

2.6   Removal of Radionuclides

39 Each of the sets of criteria governing waste determinations contains a requirement that certain
40 radionuclides be removed to the maximum extent practical. As discussed in Sections 2.4.2 and
41 2.4.3, there are differences between the radionuclide removal requirements of the NDAA, DOE
42 Order 435.1, and the West Valley Policy Statement. NDAA Criterion 2 requires that highly
43 radioactive radionuclides be removed to the maximum extent practical. DOE M 435.1-1 (DOE,
44 2001b) and the West Valley Policy Statement (NRC, 2002) require that key radionuclides be
45 removed to the maximum extent that is technically and economically practical. As discussed in
46 Section 2.4.3, NRC staff believes that NDAA wording provides more flexibility in demonstrating
47 compliance with this criterion. See Section 3 for detailed guidance on determining whether this
48 criterion has been met.

2.7 Compliance With Performance Objectives of 10 CFR Part 61, Subpart C

As discussed in Section 2.4.5, there are differences between the NDAA, DOE M 435.1-1, and the West Valley Policy Statement requirements regarding the performance objectives of 10 CFR Part 61. In most cases, the reviewer should evaluate compliance with the performance objectives in 10 CFR Part 61, Subpart C.

10 CFR 61.40 requires that land disposal facilities be sited, designed, operated, closed, and controlled after closure so that reasonable assurance exists that exposures to humans are within the limits established in the performance objectives in 10 CFR 61.41–61.44. The reviewer should evaluate compliance with this requirement by performing the reviews described in Sections 4–7 of this guidance document.

To evaluate compliance with the performance objective for the protection of the general population from releases of radioactivity (10 CFR 61.41), the reviewer should confirm that concentrations of radioactive material that may be released to the general environment in groundwater, surface water, air, soil, plants, or animals will not result in an annual dose to a member of the public that is greater than 0.25 mSv [25 mrem] and will be maintained as low as is reasonably achievable (ALARA). The reviewer should evaluate compliance with this requirement as described in Section 4 of this guidance document. Note that although 10 CFR 61.41 requires that materials released to the general environment will not result in an annual dose exceeding an equivalent of 25 millirems to the whole body, 75 millirems to the thyroid, and 25 millirems to any other organ of any member of the public, the NRC staff uses an exposure limit of 0.25 mSv [25 mrem] total effective dose equivalent (TEDE) in making this assessment (NRC, 1999, 2005). See Section 4.6.1.3 for more detailed guidance about dose calculations.

The performance objective for protection of individuals from inadvertent intrusion (10 CFR 61.42) requires that the design, operation, and closure of the land disposal facility will ensure protection of any individual inadvertently intruding into the disposal site and occupying the site or contacting the waste at any time after active institutional controls over the disposal site are removed. The performance objective does not provide numerical dose criteria for protection for the inadvertent intruder. However, NRC typically applies a whole body-dose equivalent limit of 5 mSv/yr [500 mrem/yr], as described in the Draft Environmental Impact Statement for 10 CFR Part 61 (NRC, 1981), to assess compliance with 10 CFR 61.42. The reviewer should evaluate compliance with this requirement as described in Section 5 of this guidance document.

The performance objective for the protection of individuals during operations (10 CFR 61.43) requires that land disposal facility operations will comply with the standards for radiation protection set out in 10 CFR Part 20, except for releases of radioactivity in effluents from the land disposal facility, which will be governed by 10 CFR 61.41. In addition, the performance objective requires that radiation exposures during operations are maintained ALARA. The information reviewed using Section 6 of this guidance document supports the evaluation of compliance with this requirement.

The performance objective for stability of the disposal site after closure (10 CFR 61.44) requires that a disposal facility be sited, designed, used, operated, and closed to achieve long-term stability of the disposal site and to eliminate, to the extent practicable, the need for ongoing active maintenance of the disposal site following closure so that only surveillance, monitoring, or minor custodial care is required. Evaluation of compliance with 10 CFR 61.44 is limited to a review of site stability, as described in Section 7. However, because the stability of a disposal site is important to its long-term performance, the reviewer should ensure that the effects of site instabilities identified in this part of the review are adequately modeled or bounded in the performance assessment (see Section 4) and inadvertent intruder analysis (see Section 5).

2.8 References

Public Law 108-375, "Ronald W. Reagan National Defense Authorization Act for Fiscal Year 2005." October 2004.

U.S. Department of Energy (DOE). "Radioactive Waste Management." DOE O 435.1. 2001a.

————. "Radioactive Waste Management Manual." DOE M 435.1-1. 2001b.

U.S. Nuclear Regulatory Commission (NRC). "Draft Environmental Impact Statement on 10 CFR Part 61, Licensing Requirements for Land Disposal of Radioactive Waste." NUREG–0782. September 1981.

————. "Disposal of High-Level Radioactive Wastes in a Proposed Geological Repository at Yucca Mountain, Nevada." Proposed Rule. *Federal Register*, 64 FR 8640. February 1999.

————. "Decommissioning Criteria for the West Valley Demonstration Project (M–32) at the West Valley Site; Final Policy Statement." *Federal Register*, 67 FR 5003. February 2002.

————. "NRC Review of Idaho Nuclear Technology and Engineering Center Draft Waste Incidental to Reprocessing Determination for Tank Farm Facility Residuals—Conclusions and Recommendations." Letter from L. Kokajko to J. Case, DOE. June 2003.

————. "Technical Evaluation Report for the U.S. Department of Energy, Savannah River Site Draft Section 3116 Waste Determination for Salt Waste Disposal." Letter from L. Camper to C. Anderson, DOE. December 2005.

3 RADIONUCLIDE REMOVAL AND CONCENTRATION LIMITS

As discussed in Sections 2.4.2 and 2.4.3, each set of incidental waste criteria requires that highly radioactive radionuclides be removed to the maximum extent practical. Because relevant wastes may have a wide range of physical, chemical, and radiological characteristics, the list of highly radioactive radionuclides is expected to vary among different waste determinations. In general, NRC staff believes that highly radioactive radionuclides are those radionuclides that contribute most significantly to risk to the public, workers, and the environment (NRC, 2005).

Depending on the circumstances, the removal of radionuclides may refer to the minimization of the volume of residual waste remaining in place (e.g., removal of waste from a tank that is to be closed) or to the removal of radionuclides from a waste stream (e.g., removal of radionuclides from salt waste being disposed of as saltstone at SRS). Essentially, the common goal of the various radionuclide removal criteria is to ensure that DOE minimizes inventory of highly radioactive radionuclides in wastes that are classified as incidental.

To evaluate compliance with each of the radionuclide removal criteria, reviewers should complete the following three steps:

- Evaluate the waste inventory, including sampling, analysis, calculations, and uncertainties that may affect estimates of waste volumes and radionuclide concentrations (see Section 3.1).

- Evaluate DOE's basis for selecting highly radioactive radionuclides (see Section 3.1).

- Evaluate the technical basis for determining whether radionuclides have been (or will be) removed to the maximum extent practical by (1) evaluating DOE's selection of waste removal technologies, (2) evaluating DOE's criteria for concluding that waste removal activities should be stopped, and (3) evaluating the costs and benefits of additional waste removal (see Sections 3.3 and 3.4).

This section provides review areas and procedures for each of these steps. In addition to the radionuclide removal criterion specific to the evaluation of incidental waste at the West Valley site, other radionuclide removal activities (e.g., removal of contaminated buildings) may be subject to separate analyses to support NRC license modification or termination for the site as a whole. Such site-wide analyses are not addressed in this guidance.

As discussed in Section 2.4.4, the NDAA and DOE M 435.1-1 require that wastes be classified as a function of radionuclide concentrations; the NDAA requires classification according to the concentration limits described in 10 CFR 61.55, while DOE Order 435.1 allows classification either by the concentration limits described in 10 CFR 61.55 or by alternate DOE-authorized requirements. Because the assessment of radionuclide concentrations is part of the assessment of radionuclide inventory and is essential to waste classification, this section also addresses the review of compliance with the NDAA and DOE M 435.1-1 requirements related to radionuclide concentrations and waste classification.

3.1 Inventory of Radionuclides in Waste

Evaluating the inventory of radionuclides in the waste is the first step in identifying highly radioactive radionuclides and the extent to which they can be removed. Radionuclide inventories also are needed to develop a source term model for a performance assessment and inadvertent intruder analysis (see Section 4.3.3.1.3).

3.1.1 Areas of Review

The reviewer should evaluate the description of the radionuclide inventory in the waste, including radionuclide activities and waste volumes. Radionuclides with relatively high solubility, low sorption, high dose conversion factors, and/or significant ingrowth are of particular significance.

The reviewer should evaluate the history of the wastes of interest, including information about the generation of the waste streams identified in the waste determination and any treatment processes (e.g., neutralization of acidic waste) that could influence the physical, chemical, or radiological characteristics of the waste. The reviewer also should evaluate previous inventory estimates and the reasons for any differences between the previous estimates and estimates provided in the waste determination.

The reviewer should evaluate the sampling methodologies used to determine radionuclide activities, including the rationale for the sampling approach, the particular sampling points chosen, and the justification for the number of samples collected. The reviewer should evaluate the spatial distribution of the radionuclides within the waste, including, if appropriate, both the areal and vertical distribution of radionuclides. The reviewer should ensure that the sampling methodology appropriately accounted for the heterogeneity of the waste. In assessing the representativeness of samples, the reviewer should consider any reasons why wastes of different composition may not have been adequately sampled (e.g., if the range of sample depths in a tank is limited and wastes of different composition may have stratified). The reviewer should evaluate the potential effects of access limitations (e.g., waste inside of abandoned equipment or in sand pads, limited available risers or limited possibilities to create new sampling ports in the tops of tanks) in limiting sample locations. In addition to evaluating uncertainties due to sample variability (e.g., quantified as the variability among sample measurements), the reviewer should evaluate additional uncertainty in radionuclide activities that would result if the samples do not adequately represent waste heterogeneity. In assessing waste heterogeneity, the reviewer should consider information about activities performed to mix the waste and about any areas that may have remained unmixed.

The reviewer should evaluate the analytical methods used to characterize the radionuclide inventory of the waste, including information about the calibration and sensitivity of the instruments used to make the measurements. Any assumptions made and reported uncertainties should be reviewed as well. The reviewer should evaluate the expected impact of analytic uncertainty.

DOE may estimate concentrations of radionuclides that were not measured. Inventory estimates may be based on historical knowledge (e.g., tracking of waste added to tanks) or other types of process knowledge (e.g., waste treatment efficiencies or ORIGEN2 calculations based on cladding and fuel types, burnup levels, cooling times, and other parameters). The reviewer should evaluate parameter values and assumptions used in making these estimates. The applicability of process knowledge estimates to the waste conditions should be evaluated. For example, results of ORIGEN2 calculations to estimate radionuclide inventories may not be applicable to waste in tanks in which radionuclides may have been removed to different extents because of association with different physical phases. In general, inventory estimates based on historical or process knowledge and special calculations are expected to be more uncertain than estimates based on sample measurements, and the reviewer should evaluate the technical basis for uncertainties in the estimated values or the technical basis for concluding that the estimated values are conservative. If possible, the reviewer should compare the results of similar calculations (e.g., estimates based on historical information, special calculations, or process knowledge) to one another. If possible, the reviewer should also compare the results of calculated estimates with sampled values to assess the reliability of the estimated values. Reviewers should consider whether the results of a comparison made for one radionuclide can provide information about the expected accuracy of the estimated inventory of chemically similar radionuclides. Reviewers should determine whether there is an adequate technical basis for the accuracy of estimates based on historical or process knowledge.

The reviewer should examine the methods used to determine the volume of waste used in inventory calculations (e.g., use of geostatistical methods to calculate waste volumes based on camera recordings of waste in tanks) and evaluate reported uncertainties in waste volumes. The reviewer should evaluate the expected impact of volume uncertainties on the results of the performance assessment.

3.1.2   Review Procedures

Radionuclide inventories support the evaluation of the removal of highly radioactive radionuclides as well as the development of a source-term model (see Section 4.3.3.1.3). In general, reviewers should evaluate information about waste history, sampling, measured radionuclide activities, waste volumes, and calculations performed to estimate radionuclide inventories, as well as the technical bases for the resulting radionuclide inventory estimates. It is particularly important for reviewers to evaluate the technical bases for uncertainties in inventory associated with sample heterogeneity or the use of historical or process knowledge and to determine whether uncertainties in radionuclide inventories have been adequately represented or bounded. Specifically, the reviewer should perform the following activities:

- Evaluate information about waste generation and treatment activities, and determine whether the reported radionuclide concentrations appear to be consistent with expected concentrations in contributing waste streams. For example, much lower concentrations of fission products (e.g., Sr-90, Tc-99, Cs-137) and elements used in fuel cladding (e.g., Zr, Al) would be expected in tanks receiving primarily second- or third-cycle wastes as compared to tanks receiving primarily first-cycle extraction waste.

- Verify that the predicted physical and chemical forms of radionuclides are consistent with the properties of contributing waste streams and previous waste management activities (e.g., precipitation of radionuclides during neutralization of acidic waste, partitioning of radionuclides onto zeolites).

- Evaluate previous inventory estimates, and verify that the technical bases for any differences between historical and current estimates are adequate.

- Determine whether waste samples adequately represent heterogeneity of the waste. Specifically evaluate the number, location, depth, and physical phases of samples as well as the expected degree of mixing of the waste. Identify any unsampled wastes (e.g., waste in abandoned equipment in tanks, waste in unsampled hardened mounds) and determine whether the unsampled wastes may contribute significant uncertainty to total radionuclide inventories. Verify that uncertainties resulting from waste heterogeneity are adequately represented or bounded in the reported radionuclide inventories.

- If applicable, evaluate sampling and analysis plans and data quality assessments, and verify that relevant data quality objectives are met. If the data quality objectives are not met, evaluate DOE's technical basis for concluding that the data are sufficient to support the inventory estimates (e.g., a comparison of sampling results to previous sampling results that met data quality objectives or predictions based on other techniques).

- Verify that estimates of the number of required samples provided in the sampling plan are based on accurate assumptions about the heterogeneity of the waste. For example, verify that calculations of the number of samples needed to provide the desired statistical power or upper confidence limits are not based on the assumption that the waste is well mixed if it is not well mixed.

- Assess the technical basis for any identified limitations in the number or locations of samples (e.g., limited number of sampling ports or internal obstructions in tanks, difficulties in sampling specific phases of waste, significant worker hazards), and confirm that the resulting uncertainties in total inventory have been adequately represented or bounded.

- Evaluate the analytical methods used to measure radionuclide concentrations, verify that the analytical techniques used have the appropriate sensitivity for the radionuclides of interest, and verify that the uncertainties in the analytical methods (e.g., due to instrument calibration) are either propagated into the inventory estimates or have been adequately bounded.

- Verify that DOE has adequately described sampling and analytical uncertainties and properly identified the uncertainty propagated into calculations of waste inventory (e.g., ensure that analytic uncertainty is not substituted for sample variability).

- Evaluate the statistical metric of radionuclide concentrations used to calculate inventories in the waste determination (e.g., mean, 95-percent upper confidence limit) to ensure that the technical basis for the selection is adequate and the metric is properly calculated.

- Evaluate methods used to estimate the inventory of highly radioactive radionuclides, and determine whether reasonable efforts have been made to base concentrations for highly radioactive nuclides on sample measurements. If the concentrations of highly radioactive nuclides are based on calculations or process knowledge, verify that the technical justification for using calculated rather than sampled values is adequate (e.g., lack of analytical techniques with appropriate detection limits).

- If techniques other than site-specific measurements are used to estimate radionuclide concentrations (e.g., estimates based on historical knowledge or ORIGEN2 calculations), verify that the techniques have been appropriately verified and validated. For example, compare any estimates made prior to sampling with sampled values, and verify that the estimated values were accurate or conservative or that the reasons for any nonconservative estimates are understood. Ensure that comparisons of predicted and measured values provide an adequate basis for the reliability of estimates of chemically similar radionuclides that have not been sampled.

- If estimates are based on historical data (e.g., tracking of wastes added to a waste tank) or process knowledge, ensure that significant sources of uncertainty (e.g., waste stream variability, unknown waste streams) have been adequately characterized and propagated in inventory estimates.

- If estimates are based on process knowledge (e.g., ORIGEN2 calculations), verify that the effects of subsequent waste management activities (e.g., precipitation of radionuclides during neutralization of acidic waste and subsequent incongruent removal of waste phases) are included appropriately in inventory estimates.

- Confirm that DOE has an adequate technical basis for estimating the volume of the waste. Evaluate the methods (e.g., geometry, models, statistical techniques) used for estimating waste volumes, and confirm that uncertainty has been considered and propagated into the inventory estimate. For example, if DOE uses geostatistical methods to estimate tank waste residual volumes after radionuclide removal, verify that the uncertainty resulting from the geostatistical estimate is adequately reflected in the analysis.

3.2   Identification of Highly Radioactive Radionuclides

As discussed in Section 2.4.2, the NRC staff believes that highly radioactive radionuclides are those radionuclides that contribute most significantly to risk to the public, workers, and the environment (NRC, 2005). Because highly radioactive radionuclides are defined in terms of the risk they pose to various receptors, the identification of highly radioactive radionuclides is sensitive to changes made in the performance assessment, inadvertent intruder analysis, and calculations of worker risk. Therefore, the choice of highly radioactive radionuclides should be sufficiently conservative so that all potential highly radioactive radionuclides are included, or the selection process should be iterative so that a radionuclide is added to the list if changes in the relevant risk calculations increase the predicted risk significance of the radionuclide.

A list of highly radioactive radionuclides is expected to be specific to a particular waste determination, and radionuclides that are identified as highly radioactive radionuclides in one waste determination are not necessarily expected to be identified as highly radioactive radionuclides in another waste determination.

3.2.1   Areas of Review

Because highly radioactive radionuclides are defined in terms of the risks they pose to various receptors (see Section 2.4.2), the selected highly radioactive radionuclides ultimately must be compared to the results of risk calculations. In general, the reviewer should evaluate the

1   contribution of radionuclides present in the waste to radiological dose to members of the
2   general public (including inadvertent intruders) and workers. These exposures are expected to
3   occur through different pathways and at different times (see Example 1). In reviewing the
4   identification of highly radioactive radionuclides, it is particularly important that the reviewer
5   evaluate the potential uncertainties in predicted receptor doses (see Example 2). Potential
6   contributions to uncertainties in performance assessment and inadvertent intruder results are
7   discussed in Sections 4 and 5, respectively. In identifying uncertainties, the reviewer should
8   consider the results of independent performance assessment and inadvertent intruder analyses.
9
10 DOE may identify highly radioactive radionuclides by starting with radionuclide inventories and
11 eliminating radionuclides from the list of potential highly radioactive radionuclides based on
12 screening criteria. In these cases, the reviewer should pay particular attention to uncertainties
13 in radionuclide inventories (see Section 3.1) and assess the reasonableness of any screening
14 criteria used to remove radionuclides from the list of potential highly radioactive radionuclides.
15
16 DOE may use additional criteria to identify highly radioactive radionuclides. For example, in its
17 draft waste determination for closure of the Idaho Nuclear Technology and Engineering Center
18 Tank Farm Facility, at INL (DOE, 2005), DOE included all radionuclides in Table 1 of
19 10 CFR 61.55 as highly radioactive radionuclides. In general, reviewers should focus on
20 ensuring that all risk-significant radionuclides are included as highly radioactive radionuclides,
21 and they do not need to be concerned if DOE includes additional radionuclides as highly
22 radioactive radionuclides. When assessing whether highly radioactive radionuclides have been
23 removed to the maximum extent practical, the reviewer should focus on the removal achieved
24 for risk-significant radionuclides.
25
26 Example 1
27
28 *DOE estimates that preparing a certain tank for closure by removing waste from the tank and*
29 *stabilizing the waste heel will result in a worker dose of 1 person-rem, primarily due to Cs-137.*
30
31 *In its performance assessment, DOE estimates that a hypothetical receptor that lives 100 m*
32 *[330 ft] from the tank farm and uses groundwater as a drinking water source will receive a peak*
33 *annual dose of 0.03 mSv [3 mrem] at 3,000 years after the end of active institutional controls.*
34 *Of this annual dose, DOE estimates 0.02 mSv [2 mrem] is due to ingestion of I-129, 0.005 mSv*
35 *[0.5 mrem] is due to ingestion of Se-79, 0.004 mSv [0.4 mrem] is due ingestion of Tc-99, and*
36 *0.001 mSv [0.1 mrem] is due to contributions of other radionuclides. In addition, DOE also*
37 *predicts that a hypothetical receptor living in the same location 5,000 years after site closure will*
38 *receive an annual dose of 0.015 mSv [1.5 mrem] due primarily to ingestion of Np-237.*
39
40 *DOE also predicts that an individual who drills through a tank and is inadvertently exposed to*
41 *waste will receive a peak dose of 2 mSv [200 mrem] 1 year after site closure. DOE predicts that*
42 *1 mSv [100 mrem] of this dose is attributable to direct radiation by Sn-126, 0.6 mSv [60 mrem]*
43 *is due to inhalation of Pu-239, 0.3 mSv [30 mrem] is attributable to inhalation of U-235, and*
44 *0.1 mSv [10 mrem] is attributable to other radionuclides.*
45
46 *In this case, Cs-137, I-129, Se-79, Tc-99, Np-237, Sn-126, Pu-239, and U-235 would be*
47 *identified as highly radioactive radionuclides for this waste determination.*
48

Example 2

DOE plans to close a waste tank after removing waste from the tank and stabilizing the waste heel with grout. DOE estimates that these activities will result in a worker dose of 1 person-rem, almost entirely due to Cs-137. In its performance assessment, DOE estimates that a hypothetical receptor that lives 100 m [330 ft] from the tank farm and uses groundwater as a drinking water source will receive a peak annual dose of 0.015 mSv [1.5 mrem], almost entirely due to Np-237, 5,000 years after site closure. DOE also estimates that an individual who drills through a tank and is inadvertently exposed to waste will receive a peak dose of 1 mSv [100 mrem], almost entirely due to direct radiation by Sn-126. Therefore, DOE's original list of highly radioactive radionuclides for the waste determination includes only Cs-137, Np-237, and Sn-126.

In a sensitivity analysis performed during NRC's review, DOE uses an alternate conceptual model to predict that a hypothetical receptor living 100 m [330 ft] from the tank farm who uses groundwater as a drinking water source will receive a peak annual dose of 0.02 mSv [2 mrem], almost entirely due to Tc-99, 2,000 years after site closure. In this case, if the alternate conceptual model does not represent a very unlikely scenario, Tc-99 should be added to DOE's list of highly radioactive radionuclides for the waste determination.

3.2.2   Review Procedures

The reviewer should ensure that initial identification of highly radioactive radionuclides is sufficiently conservative that it does not omit radionuclides that may be predicted to cause a significant contribution to risk after uncertainties in risk calculations are resolved. The reviewer should then evaluate any iterative analysis that DOE uses to refine the initial list, identify those radionuclides that contribute most to dose, and identify those uncertainties that are expected to have the most significant affect on predicted dose. Specifically, the reviewer should perform the following review procedures.

- If a screening analysis is used

    — Ensure that the initial list of radionuclides included in the analysis is comprehensive;

    — Evaluate screening criteria DOE used to eliminate potential highly radioactive radionuclides (e.g., short half-life, low total activity in the waste), and verify that they are reasonable and sufficiently conservative; and

    — Ensure that ingrowth of daughters is considered and that parent radionuclides are not inappropriately screened from analysis.

- Verify that DOE has considered the appropriate receptors (i.e., workers and members of the public including inadvertent intruders) in defining its list of highly radioactive radionuclides.

- Ensure that uncertainties in radionuclide inventories are adequately represented in initial lists of radionuclides used in screening procedures (see Section 3.1).

- Ensure that uncertainties in performance assessment and inadvertent intruder analyses are adequately represented in the selection of highly radioactive radionuclides (see Sections 4 and 5).

- Use independent analyses to assess which radionuclides are expected to cause the most significant risk to members of the public, including the inadvertent intruders. Vary model parameters and assumptions to evaluate a reasonable range of alternative scenarios (e.g., regarding barrier performance, biointrusion, flooding, or seismic events). Examine the reasons for any differences between DOE's list of highly radioactive radionuclides and the radionuclides identified as causing significant risk in the independent analyses, and request additional technical basis supporting the DOE analysis, as necessary.

3.3 Radionuclide Removal Technology Selection and Implementation

As discussed in Section 2.4.3, "removal" of radionuclides refers to both removal of waste from a disposal system (either by removing waste alone or by removing contaminated structures that contain waste) and treatment of waste to remove radionuclides from the waste stream. As discussed in Sections 2.4.2 and 2.4.3, there are differences between the NDAA, DOE Order 435.1 (DOE, 2001), and the West Valley Policy Statement (NRC, 2002) requirements regarding the extent of removal of radionuclides, and NRC staff believes the NDAA wording allows for broader considerations in demonstrating compliance with this criterion. Typically, however, reviewers should expect to evaluate a quantitative analysis that demonstrates that radionuclides have been removed to the maximum extent practical. For example, impacts of additional removal activities on removal schedules may be related to cost or worker risk.

Because DOE may submit a waste determination either before (DOE, 2005a) or after (DOE, 2005b) it has stopped relevant removal operations, evaluation of compliance with the radionuclide removal criterion may be based on either the extent of removal DOE has performed or the removal that DOE indicates it will achieve before final disposal actions are taken. Thus, the reviewer may conclude either that highly radioactive radionuclides "have been" or "will be" removed to the maximum extent practical. If DOE submits a waste determination for INL or SRS prior to ending removal actions, NRC staff will expect to monitor the extent of radionuclide removal achieved and assess any impacts on meeting the performance objectives of 10 CFR Part 61, Subpart C (Section 10), including the ALARA requirement of 10 CFR 61.41 (see Section 4.7). If removal activities are not completed as described in DOE's waste determination, conclusions NRC staff made regarding radionuclide removal to the maximum extent practical that were based on the waste determination review may no longer be valid. However, NRC will not reassess its conclusions about radionuclide removal or compliance with NDAA Criterion 2 during monitoring, because NRC's monitoring role under the NDAA is limited to assessing compliance with the performance objectives of 10 CFR 61, Subpart C.

In general, DOE's technology selection process is expected to focus on relatively mature technologies. However, less mature technologies may be relevant if DOE proposes a multistep removal procedure in which some technologies may not be implemented for several years after DOE's final waste determination. Reviewers also may want to note less mature technologies they encounter during their reviews for possible consideration during subsequent reviews.

3.3.1 Areas of Review

As discussed in Section 2.4.3, an NRC reviewer is expected to evaluate both technical and economic aspects of radionuclide removal. In general, review of the technical aspects of radionuclide removal is expected to include an assessment of the technologies DOE selected to perform radionuclide removal, while review of the economics of additional removal is expected to include an assessment of the costs and benefits of additional radionuclide removal (see Section 3.4). Although reviews of technology selection and cost-benefit analyses are expected to be performed for each waste determination review, some components of the review may depend on the timing of the submission of the waste determination with respect to the status of removal actions. Specifically, if DOE stops relevant removal operations before submitting a waste determination, the review should include an evaluation of the removal actions that have occurred and the decision to stop removal activities, while reviews of waste determinations submitted prior to stopping radionuclide removal operations should include an evaluation of the criteria DOE establishes to determine when removal is complete.

In reviewing the selection of radionuclide removal technologies, the reviewer should evaluate the decision process DOE used to select appropriate technologies as well as the final technology selections. Reviewers should evaluate the range of technologies DOE considered, the sources of information DOE used, and how information was used in DOE's technology selection process. Information about available technologies may be found in reports generated by other DOE sites, reports from third parties [e.g., the National Academy of Sciences (NAS) or Defense Nuclear Facilities Safety Board (DNFSB)], and reports from industrial or international sources. In addition to evaluating the range of technologies DOE considered, reviewers should evaluate whether DOE considered potentially beneficial sequencing of and synergisms among different technologies.

In assessing removal actions that have been stopped, the reviewer should evaluate the selection process based on the information and technologies available at the time of selection. In assessing the practicality of additional radionuclide removal, the reviewer should assess technologies available at the time of the review, including technologies that have been used successfully at other sites are of particular interest. Reviewers should evaluate DOE's rationale for selection of waste removal techniques, including a comparative assessment of the technical and economic characteristics of the techniques considered.

NRC staff believes that, in many cases, the intent of requiring removal of highly radioactive radionuclides to the maximum extent practical can be satisfied by reducing the volume of residual waste in a contaminated structure (e.g., a tank, an evaporator) to the maximum extent practical. However, as discussed in Section 2.4.3, reviewers also should consider (1) whether it would be practical to remove selected highly radioactive radionuclides from the waste (e.g., by chemical extraction) or (2) whether it would be practical to remove the contaminated structure for disposal instead of stabilizing it and disposing of it in place. As discussed in Section 3.4, in some cases a relatively simple analysis may demonstrate that an alternative is not practical.

To the extent practical, reviewers should consider the impacts of the proposed waste management activities and alternative waste management activities on the broader waste management system. For example, if generating additional liquid waste presents technical challenges because of tank space limitations, reviewers should consider whether alternate technologies would generate less secondary liquid waste than the proposed technologies

1 (e.g., by recycling liquids rather than using clean water). Reviewers also should consider any
2 physical or chemical impacts of removal methods on downstream processes (e.g., effects of
3 oxalic acid on evaporators or vitrification equipment).
4
5 In addition to reviewing DOE's technology selection process, reviewers should evaluate the
6 technical basis for stopping removal activities. In general, the decision to terminate removal
7 activities should be based on a demonstration that additional removal would be impractical. For
8 example, a statement that 99 percent of waste has been removed from a tank is not a sufficient
9 basis for stopping removal, but a demonstration that removing the remaining waste would not
10 substantially reduce risks and would cause excessive worker dose would be sufficient. In
11 assessing the basis for stopping removal activities, the reviewer should evaluate available
12 documentation of removal goals that were established before radionuclide removal waste
13 started. The reviewer should evaluate both the technical basis for the original removal goal and
14 the consistency of the removal achieved with the removal goal. The actual extent of removal
15 achieved in terms of radionuclide inventories and the uncertainties in those inventories should
16 be reviewed as described in Section 3.1.
17
18 Although removal fractions alone are not expected to provide a basis for terminating removal
19 activities, they may help a reviewer assess the success of competing removal technologies. If
20 removal efficiencies are used to compare the success of competing technologies, reviewers
21 should ensure efficiencies are based on comparable starting points. For example, with respect
22 to removal of waste from tanks, reviewers should determine whether starting inventories were
23 based on the historical maximum radionuclide inventories, inventories after bulk removal, or
24 another baseline inventory. Comparisons are expected to be useful only if consistent baselines
25 can be established. Reviewers should note that the distinction between "bulk" and "heel"
26 removal typically is subjective.
27
28 If removal activities are stopped before removal goals are reached or if removal activities have
29 been stopped and no information about removal goals is available, reviewers should assess
30 DOE's basis for stopping removal. For example, if DOE indicates that radionuclide removal was
31 stopped because of declining removal efficiency (e.g., decreasing amount of waste removed
32 from a tank per mixing cycle), the reviewer should evaluate whether DOE considered
33 opportunities to improve effectiveness (e.g., by redirecting mixing jets). The reviewer also
34 should evaluate whether continued removal at the decreased efficiency still produces significant
35 removal. For example, removal efficiency may have declined from 5,000 liters [1,300 gal] to
36 1,000 liters [260 gal] of waste per mixing cycle, but removal of 1,000 liters [260 gal] of waste per
37 mixing cycle may still be practical. If DOE indicates that waste removal was stopped because of
38 equipment failure, the reviewer should investigate the costs and other impacts (e.g., on
39 schedule) of replacing the failed equipment. In general, if DOE identifies schedule constraints
40 as a reason why waste removal actions are stopped or why additional removal actions cannot
41 be completed, reviewers should assess the implications of the identified schedule impacts on
42 more quantitative metrics such as cost in comparison to worker risk.
43
44 If a waste determination is submitted prior to completion (or commencement) of waste removal
45 activities, the reviewer should evaluate DOE's criteria for determining when waste removal is
46 complete. Essentially, a reviewer should evaluate the technical basis for concluding DOE will
47 meet its removal goals and the practicality of achieving a greater extent of radionuclide removal
48 than required by the DOE proposed criteria. For example, if DOE proposes that removal will be
49 complete if a certain volume of waste has been removed from a tank, the reviewer should

1   evaluate DOE's technical basis for concluding the specified volume of waste will be removed
2   and the practicality of removing a greater volume of waste from the tank. Similarly, if DOE
3   proposes to remove radionuclides from a waste stream with a chemical treatment process, the
4   reviewer should evaluate the technical basis for the predicted effectiveness of removal of
5   radionuclides from the waste stream as well as the practicality of accomplishing additional
6   radionuclide removal.
7
8   For removal activities that have not been completed, the reviewer also should identify the main
9   factors that could cause changes in the proposed approach. For example, a technology that is
10   less mature may introduce greater uncertainty in the extent of radionuclide removal that can be
11   accomplished or increase the cost of the proposed waste management activities as compared
12   to more mature technologies. To assess technological maturity, the reviewer should assess
13   information about technologies that have been used or developed at other DOE sites as well as
14   reports from third parties (e.g., NAS or DNFSB). If DOE proposes a technology that has not yet
15   been used successfully under similar situations, the reviewer should evaluate DOE's plans for
16   changing the waste management approach to accommodate unforeseen problems (e.g., DOE's
17   plans for using alternate technologies to complete radionuclide removal if the originally selected
18   technology cannot achieve the removal goals). Additional uncertainties in the conclusion that
19   radionuclides will be removed to the maximum extent practical may result if radionuclide
20   removal plans extend over long time periods or if removal activities are not expected to begin for
21   a significant amount of time after a waste determination is submitted (e.g., greater than 5 years).
22   These uncertainties may result because it may be difficult to conclude that the proposed
23   radionuclide removal activities will remain practical in the future (e.g., because of changing
24   programmatic goals). If waste treatment activities are expected to begin several years after the
25   waste determination has been submitted, the reviewer should evaluate DOE's process for
26   considering technological developments that occur after the submission of the waste
27   determination. In addition, reviewers must consider uncertainties in proposed waste
28   treatment schedules when evaluating the costs and benefits of a particular treatment approach,
29   because changes in the schedule could significantly affect the relative costs of various
30   treatment alternatives.
31
32   The primary benefit of radionuclide removal is expected to be a reduction in the risk the waste
33   will pose to the general public, including inadvertent intruders. Because the performance
34   assessment and inadvertent intruder analysis provide the basis for quantifying the risk the
35   waste will pose to members of the public, an analysis of the costs and benefits of additional
36   radionuclide removal is expected to depend in part on the performance assessment and
37   inadvertent intruder analysis dose predictions. In addition, radionuclide removal may affect the
38   risk to individuals during operations (see Section 6) or site stability (see Section 7). Therefore,
39   the reviewer should evaluate the consistency of information presented in a cost-benefit analysis
40   to support conclusions about the practicality of additional radionuclide removal with information
41   reviewed using Sections 4–7 of this guidance document. For example, if additional removal
42   would require changes to the physical or chemical form of the waste, the reviewer should
43   evaluate the potential impacts of the changes on site stability. Similarly, if DOE indicates that
44   additional removal would cause unacceptable worker risks, the consistency of the predicted
45   worker risks due to additional radionuclide removal with the information presented regarding
46   protection of workers during operations should be evaluated. Furthermore, if changes to the
47   performance assessment or inadvertent intruder analysis are made during the review process,
48   the reviewer should evaluate potential impacts on any cost-benefit analysis used to support the
49   conclusion that highly radioactive radionuclides have been removed to the maximum extent

1 practical. Additional information about reviewing cost-benefit analyses is provided in
2 Section 3.4.
3
4 **3.3.2   Review Procedures**
5
6 Reviewers should evaluate waste management activities that DOE has performed or proposed
7 as described in the waste determination in comparison to alternate waste management
8 activities. In addition, reviewers should evaluate the technological feasibility and practicality
9 of additional removal of radionuclides. Specifically, the reviewer should perform the
10 following assessments:
11
12 •   Review DOE's basis for radionuclide removal technology selection. Determine whether
13   a reasonable range of potential technologies was evaluated by comparing the list of
14   technologies considered to technologies used or developed at other DOE sites, in
15   industrial applications, and in the international community, and by reviewing relevant
16   documents by other organizations, if available (e.g., NAS, DNFSB). Technologies
17   considered should include, but not necessarily be limited to, sluicing (e.g., Schlahta and
18   Brouns, 1998), mixing (e.g., Leishear, 2004; Hatchell, et al, 2001), chemical cleaning
19   (e.g., Sams, 2004), vacuum retrieval techniques (e.g., Sams, 2004), mechanical
20   manipulators (e.g, DOE, 1998; Evans, 1997), and robotic vehicles (e.g., Vesco,
21   et al., 2001).
22
23 •   As appropriate, determine whether DOE considered both methods to (1) remove waste
24   from the disposal system (either by reducing the volume of waste in a contaminated
25   structure or by removing a contaminated structure for disposal) and (2) remove
26   radionuclides from relevant waste streams (e.g., chemically extracting radionuclides
27   from waste that will remain in a tank).
28
29 •   Review DOE's documentation of its process for selecting radionuclide removal
30   technologies, and determine whether the selection process was based on appropriate
31   sources of information that were reasonably current at the time the technology selections
32   were made (e.g., expert elicitation, reports from other DOE sites, reports from industrial
33   or international sources).
34
35 •   Determine whether the criteria DOE used to select radionuclide removal technologies
36   are reasonable. In general, reviewers should expect that selection criteria include an
37   estimate of the likelihood of achieving removal goals with the selected technology, the
38   technological uncertainties associated with each technology (e.g., technological
39   maturity), the costs of implementing the technology, and an evaluation of the
40   system-wide impacts of using the technology (e.g., chemical effects on downstream
41   systems, generation of secondary waste streams requiring storage).
42
43 •   If waste treatment activities are expected to begin several years after the waste
44   determination has been submitted, evaluate DOE's process for considering
45   technological developments that occur after the submission of the waste determination.
46   Determine whether the process will require evaluation of an appropriate range of
47   alternative technologies and will allow DOE to assess whether it would be practical to
48   improve radionuclide removal by implementing a technology that was developed or
49   improved after submission of the waste determination.

- Examine DOE's documentation of radionuclide removal activities that have been stopped, and verify that key reasons used to support the termination of removal operations (e.g., availability of equipment, changes in equipment performance, programmatic considerations) were reasonable. For example, if DOE cites equipment deterioration as a reason for stopping removal, evaluate efforts to replace failing equipment or improve equipment performance.

- Determine whether decisions to stop waste removal activities based on programmatic or schedule constraints can be supported by an evaluation of the costs and potential benefits of continuing removal operations.

- If predicted doses to members of the public (including inadvertent intruders) were used to support the decision to stop radionuclide removal activities, determine whether uncertainties in the dose estimates would significantly affect the metrics used to support the decision to stop removal (e.g., cost per averted dose).

- Identify any removal goals DOE established before radionuclide removal began, and determine whether the goals have been met. If the goals have been met, evaluate the technical basis for the goals, and determine whether the goals were consistent with radionuclide removal to the maximum extent practical (i.e., determine whether significantly more radionuclide removal than required by the removal goals would have been practical).

- If removal goals DOE identified prior to radionuclide removal efforts have not been met, determine whether the technical basis for stopping removal prior to achieving the removal goals is adequate. For example, if technological limitations are cited as the reason goals were not met, the reviewer should determine why waste removal was more difficult than originally predicted (e.g., because of poor equipment performance or unforeseen characteristics of the waste).

- If DOE based its decision to terminate removal activities on declining removal efficiency, verify that reported removal efficiencies are reasonably reliable. For example, if detectors are placed on the outside of piping to measure the activity of waste being removed from a tank, verify that the detectors have the appropriate sensitivity when placed outside of the pipe to provide a reliable estimate of waste removal.

- If DOE based its decision to terminate removal activities on declining removal efficiency, determine whether DOE considered whether modifications to the removal process (e.g., redirection of mixing jets) could improve removal.

- If DOE based its decision to terminate removal activities on declining removal efficiency, determine whether DOE considered whether it would be practical to implement alternate technologies or methods to improve removal.

- If removal operations have not been completed, verify that DOE's proposed criteria for terminating removal operations include minimum standards that are consistent with meeting the performance objectives. For example, if DOE will conclude tank waste removal is complete when the removal rate decreases to a certain value, verify that the criteria also contain minimum removal amounts such that the inventories of highly radioactive radionuclides left in the tank (including uncertainties) will be consistent with or less than the source term used in the performance assessment and inadvertent intruder analysis (see Section 4.3.3.1.3).

- If removal operations have not been completed, examine DOE's criteria for terminating removal operations and determine whether additional removal would be practical. For example, if removal of waste in a tank is to be considered complete when the removal rate drops to a certain value, determine whether continuing operation would be expected to result in a cost-effective reduction in dose to a member of the public (see Section 3.4).

- If removal operations have not been completed, verify that predicted removal efficiencies for proposed removal operations are supported by a sufficient technical basis. For example, if waste will be treated by sorption of radionuclides onto a finely dispersed solid and subsequent filtration of that solid from the liquid waste, determine whether the reported predicted efficiencies of sorption and filtration are consistent with efficiencies observed during laboratory or pilot-scale tests under similar conditions or similar experience at DOE sites.

- Identify technological challenges DOE described as limiting factors in radionuclide removal. If a stated technological challenge is expected to recur in subsequent removal activities (e.g., if a waste with problematic physical characteristics is present in other tanks at the site), note the challenge DOE efforts to resolve it in the TER (see Section 9).

- Review any cost-benefit analysis used to support the decision that additional radionuclide removal would not be practical (see Section 3.4), and confirm that the analysis supports the conclusion that highly radioactive radionuclides have been removed to the maximum extent practical.

3.4   Cost-Benefit Analysis

Section 3.3 guides reviewers evaluating whether DOE has optimized its selection and implementation of radionuclide removal technologies to remove highly radioactive radionuclides to the maximum extent practical. This section provides additional guidance specifically focused on assessing the practicality of additional radionuclide removal, which typically is evaluated with some form of cost-benefit comparison (e.g., Gilbreath, 2005). In this guidance, the term "additional removal" indicates either more radionuclide removal than DOE already has performed, if DOE has terminated removal operations or more radionuclide removal than DOE indicates it plans to perform if DOE has based its waste determination on removal activities that are not complete. "Additional" radionuclide removal also indicates removal beyond the removal performed to meet the applicable performance objectives, which must be met irrespective of practicality.

In general, the reviewer should determine whether it is practical to reduce the inventory of radionuclides addressed in DOE's waste determination. Although the reviewer may note opportunities to optimize removal on a larger scale (e.g., by removing more waste from tanks without cooling coils so that less waste will need to be removed from tanks with cooling coils to meet the performance objectives), these decisions may be beyond the scope of an individual waste determination. The reviewer may note these observations in a TER, but should not use

1 them as a basis for determining whether highly radioactive radionuclides have been removed to
2 the maximum extent practical.
3
4 As discussed in Section 2.4.3, the NDAA allows for a broader range of factors to be considered
5 in the evaluation of the practicality of waste removal than allowed by the Manual for DOE Order
6 435.1 or the West Valley Policy Statement. However, the NRC staff believes that these factors
7 should be quantified, to the extent possible, in terms of financial costs and potential doses to
8 workers and members of the public (including inadvertent intruders) to facilitate comparison
9 of options.
10
11 **3.4.1   Areas of Review**
12
13 The reviewer should evaluate DOE's description of the potential risks, costs, and benefits
14 associated with various options for additional radionuclide removal. Costs could include, but not
15 necessarily be limited to, financial costs, delays, increases in risks to workers and members of
16 the public, system impacts (e.g., generation of secondary waste streams requiring storage in
17 tanks), and transportation risks (if waste is moved offsite for disposal). Benefits may include,
18 but not necessarily be limited to, decreases in radiological risks to workers and members of the
19 public (including inadvertent intruders), reduction in impacts on natural resources, and reduction
20 in costs of other entities incurred because of effects on natural resources.
21
22 Table 3-1 lists potential costs and benefits associated with radionuclide removal. In general, the
23 level of detail of the review should be based on the level of detail necessary to distinguish
24 between various options and to compare the practicality of additional removal with the
25 practicality of completed removal activities or other proposed waste management activities. In
26 many cases, a simplified screening analysis may be sufficient to eliminate particular waste
27 management alternatives. Therefore, it may not be necessary to consider all of the potential
28 costs and benefits listed in Table 3.1 in every case. The primary costs and benefits considered
29 in most cases are expected to be financial costs and increases or decreases in potential doses
30 to workers and members of the public.
31
32 The comparison of potential costs and benefits should be quantitative to the extent practical.
33 For example, if DOE indicates that additional radionuclide removal is impractical because it
34 would significantly delay its tank closure schedule, the reviewer should evaluate any increases
35 in dose or financial cost that DOE expects to result from the schedule delay. In reviewing the
36 waste determination for salt waste disposal at the Savannah River Site, NRC staff considered
37 the potential costs of schedule impacts, facility slowdown, and tank space issues in evaluating
38 the practicality of additional removal of highly radioactive radionuclides (NRC, 2005).
39
40 Although quantitative comparisons are easier to make than qualitative comparisons, qualitative
41 information that may help reviewers to compare various options (e.g., potential environmental
42 benefits) also should be considered. Furthermore, in some cases, qualitative differences are
43 recognized, and purely quantitative comparisons may be inappropriate. For example,
44 radiological risks to workers are not directly comparable, on a quantitative basis, to predicted
45 doses to members of the public, because worker risks are accepted risks, whereas members of
46 the public are likely to be unaware of, and may derive no benefit from, the actions that could
47 lead to a radiological dose.
48
49

Table 3-1. Potential Costs and Benefits Associated With Performing Additional Radionuclide Removal Beyond Removal Performed to Meet the Applicable Performance Objectives	
Potential Benefits	Potential Costs
Averted long–term dose to members of the public, including potential inadvertent intruders	Radiological dose to workers due to additional radionuclide removal activities
Reduction in radiological dose to workers because of increased waste stabilization, decreased numbers of waste transfers in tank farms, or other similar considerations	Financial costs of additional radionuclide removal
Decrease in costs of other entities, such as a reduction in costs incurred by public water supply utilities to meet the requirements of the Safe Drinking Water Act	Additional transportation risks
Reduction of impact on natural resources, such as groundwater aquifers	Chemical and physical effects of removal activities on downstream waste processing or storage systems
Improvement of esthetics, changes in land use, and reduction in monitoring costs*	Additional impacts on DOE's mission or schedule
	Doses to the public due to additional removal activities
	Environmental disruption due to additional removal activities
	Nonradiological workplace accidents due to additional removal activities
*Not expected to be applicable in most cases, as discussed in the text.	

In general, dose limits that are protective of human health are expected to be protective of non-human species as well [e.g., (ICRP, 1991)]. However, it is appropriate to consider any significant environmental benefit or detriment that could result from additional radionuclide removal. For example, if removing additional waste would significantly reduce environmental risks by reducing the release of radionuclides from a waste, it would be appropriate to consider that as a benefit of additional radionuclide removal. Alternately, if exhuming contaminated piping would cause environmental disruption, it would be appropriate to consider that impact as a cost of additional radionuclide removal. In addition, it is possible that releases from waste that meets the performance objectives of 10 CFR Part 61, Subpart C, could cause wellhead concentrations of some radionuclides to exceed U.S. Environmental Protection Agency's Maximum Contaminant Levels. As discussed in other contexts (e.g., NRC's guidance regarding ALARA analyses for license termination [NRC, 2003b, Appendix N]), it is appropriate to include

1   the potential reduction in costs that other entities, such as public water supply utilities, may incur
2   to meet the requirements of the Safe Drinking Water Act as benefits of additional radionuclide
3   removal in cost-benefit analyses.

4

5   Improved esthetics and changes in future land use typically are relevant to decisions between
6   alternatives that result in different potential land uses. These considerations are not expected
7   to apply to most waste determinations because (1) most relevant wastes are located among
8   industrial facilities that DOE intends to control for the foreseeable future and (2) most
9   radionuclide removal or disposal options are not expected to result in significantly different
10 possible land use scenarios. However, these considerations are listed as a potential benefit of
11 additional removal because some radionuclide removal options, such as exhuming tanks or
12 buried equipment, could permit different potential land uses than alternative options would
13 allow. Similarly, changes in monitoring costs are not expected to be relevant to most waste
14 determinations, because similar monitoring activities are expected to be appropriate following
15 most radionuclide removal options. However, reduction in monitoring costs are included as a
16 potential benefit because some radionuclide removal options could limit the need for certain
17 monitoring activities.

18

19 As previously noted, most cost-benefit analyses to support waste determinations are expected
20 to be relatively simple. However, in some cases, more detailed cost-benefit analyses may be
21 used to evaluate the practicality of a proposed approach. General guidance on cost-benefit
22 analysis is discussed in several reports (NRC, 1997; Office of Management and Budget, 1992,
23 2003). NRC also has developed cost-benefit analysis guidance for environmental reviews
24 (NRC, 2003a, Section 6.7) and for site decommissioning (NRC, 2003b, Vol. 2, Appendix N).
25 Cost-benefit assessment for a hypothetical low-level waste (LLW) disposal facility is described
26 in the Draft Environmental Impact Statement NRC prepared for 10 CFR Part 61 (NRC, 1981,
27 Vol. 2, Section 3.8). Reviewers may refer to these reports for general guidance about
28 cost-benefit analysis techniques and parameters. For example, if DOE presents a detailed
29 cost-benefit analysis that includes non-radiological accidents and transportation risks as costs
30 of additional radionuclide removal, a reviewer may compare the risks of fatalities due to
31 transportation and industrial accidents used in DOE's analysis to the values provided in
32 Table N.2 of NRC's guidance for license termination (NRC, 2003b). Although DOE may use
33 parameter values that differ from values presented in Office of Management and Budget (OMB)
34 or NRC guidance documents, reviewers may compare DOE's values to values presented in
35 these guidance documents to determine an appropriate level of detail for their evaluation of
36 DOE's parameter values. In general, reviewers are expected to perform a detailed review of
37 parameter values that significantly affect the analysis results and that differ from the values
38 presented in other NRC or OMB guidance.

39

40 Although many of the underlying principles and methods described in general guidance about
41 cost-benefit analyses may be applicable, certain specific aspects of the general guidance are
42 not expected to be applicable to waste determination reviews. Most notably, it is unclear
43 whether it is appropriate to apply specific cost-benefit metrics discussed in the general guidance
44 to DOE waste determinations. It appears to be more appropriate to compare the costs and
45 benefits of additional radionuclide removal to the costs and benefits of other similar DOE
46 risk-reduction activities (see Examples 1–4). In particular, the $2,000 per person-rem

1 conversion factor[2] that NRC uses in some contexts (e.g., regulatory analyses, ALARA analyses
2 for license termination) may not be a useful metric to apply to waste determination reviews,
3 because the metric is based on collective dose and it is designed to be applied with economic
4 discounting. The long performance period relevant to waste determinations hinders the use of
5 any metric based on collective dose because it is unrealistic to attempt to predict what the
6 population near a disposal site will be for thousands of years after site closure. In addition,
7 NRC staff previously has recommended that the monetary value associated with averted future
8 doses not be discounted in analyses relevant to 10 CFR Part 61 (NRC, 2000).
9
10 Example 1
11
12 *DOE indicates it would cost $1,000,000 to remove an additional 4,000 L [1,060 gal] of waste*
13 *from a tank and that removal would reduce the predicted 50-year dose to a member of the*
14 *public by 0.05 mSv [5 mrem]. Thus DOE calculates additional radionuclide removal would cost*
15 *approximately $20,000,000 per mSV [$200,000 per mrem] and decides not to perform additional*
16 *removal. The reviewer would compare this cost (normalized to dose reduction) to the*
17 *normalized cost DOE incurs in other similar risk-reduction activities. A finding that*
18 *$20,000,000 per mSv [$200,000 per mrem] is significantly greater than the normalized cost*
19 *DOE typically incurs in similar risk-reduction activities would support the decision that highly*
20 *radioactive radionuclides had been removed to the maximum extent practical.*
21
22 Example 2
23
24 *DOE determines it could use a chemical process to reduce the concentration of actinides in a*
25 *waste stream by 85 percent at a cost of $10,000,000. Because the release of actinides from the*
26 *final wasteform is expected to be solubility limited, the concentration reduction is not expected*
27 *to significantly affect the dose to a member of the public whose primary exposure pathway is*
28 *groundwater consumption. However, the concentration reduction is expected to reduce the*
29 *estimated dose to an inadvertent intruder from 2 mSv [200 mrem] to 0.03 mSv [30 mrem]. A*
30 *finding that $5,900,000 per mSv averted [$59,000 per mrem] is significantly greater than the*
31 *cost DOE typically incurs to avert potential doses to inadvertent intruders would support*
32 *the decision that highly radioactive radionuclides had been removed to the maximum*
33 *extent practical.*
34
35 Cost-benefit analyses are expected to be relatively simple in most cases. However, significant
36 uncertainties should be considered. The main sources of uncertainty are expected to be the
37 amount of radionuclide removal that could be accomplished by implementing a particular
38 radionuclide removal technology and the resulting reduction in dose. If significant,
39 uncertainties in the cost of implementing a particular removal technology also should be
40 included. However, in many cases, uncertainties in the predicted dose may be significantly
41 greater than cost uncertainties.
42
43

---

[2]The basis for NRC's $2,000 per person-rem metric is discussed in NUREG–1530 (NRC, 1995a).

1 <u>Example 3</u>
2
3 *DOE determines it could remove 12,000 L [3,200 gal] of additional waste from a tank at a cost*
4 *of $10,000,000. DOE also determines that removing the additional waste would reduce the*
5 *potential 50-year dose to a member of the public by 0.02 mSv [2 mrem]. Based on the*
6 *$500,000,000 cost per mSv averted [$5,000,000 per mrem], DOE determines that additional*
7 *removal would be impractical. However, revisions to the performance assessment made as a*
8 *result of the review process indicate the potential dose to a member of the public is much*
9 *greater than originally predicted, but still within the limits established by the performance*
10 *objectives of 10 CFR Part 61. As a result, DOE determines that removal of 12,000 L [3,200 gal]*
11 *would reduce the potential 50-year dose to a member of the public by 10 mSv [1,000 mrem]*
12 *{e.g., by reducing the annual dose to an individual by 0.20 mSV [20 mrem] for 50 years}. If the*
13 *resulting $1,000,000 cost per mSv averted [$10,000 per mrem] is similar to the cost per averted*
14 *dose DOE incurs in similar risk-reduction activities, the reviewer should question DOE's*
15 *assessment that additional radionuclide removal is not practical.*
16
17 <u>Example 4</u>
18
19 *DOE determines it could use a combination of a high-pressure water jet and a new type of*
20 *transfer pump to remove approximately 12,000 L [3,200 gal] of additional waste from a tank at a*
21 *cost of $2,000,000. However, DOE estimates that the new technology may remove as much as*
22 *24,000 L [6,300 gal] or as little as 6,000 L [1,600 gal] of waste from the tank. DOE estimates*
23 *that removal of 12,000 L [3,200 gal] of additional waste would result in a 2 mSV [200 mrem]*
24 *reduction in the 50-year dose to a hypothetical member of the public. After considering*
25 *uncertainties in the estimate, DOE has confidence that the dose reduction resulting from*
26 *removal of 12,000 L [3,200 gal] would be between 0.5 mSv [50 mrem] and 3 mSV [300 mrem].*
27 *DOE also estimates that the expected dose reduction would be roughly proportional to the*
28 *volume of waste removed. Therefore, DOE estimates that removal of 24,000 L [6,300 gal] of*
29 *waste would result in a potential 50-year dose reduction of 1 mSv [100 mrem] to 6 mSv [600*
30 *mrem] and that removal of 6,000 L [1,600 gal] would result in a 50-year dose reduction of 0.25*
31 *mSv [25 mrem] to 1.5 mSv [150 mrem]. The resulting estimated cost per averted dose would*
32 *range from $330,000 per mSv averted [$3,300 per mrem] to $8,000,000 per mSv averted*
33 *[$80,000 per mrem]. If this range spans the range of costs (normalized to dose reduction) DOE*
34 *incurs in similar risk-reduction activities, reviewers would need to rely on the value of the best*
35 *estimates of the normalized cost, the probability that the normalized cost will be significantly*
36 *different from the normalized costs DOE incurs to achieve similar risk-reductions, and*
37 *consideration of additional factors (such as unquantifiable factors listed in Table 3-1 and the*
38 *factors discussed in Section 3.3) to determine whether DOE has removed highly radioactive*
39 *radionuclides to the maximum extent practical.*
40
41 **3.4.2   Review Procedures**
42
43 The reviewer should evaluate the risks, costs, and benefits associated with alternative waste
44 management strategies and radionuclide removal technologies to determine whether there is
45 sufficient technical justification to conclude that waste removal is complete, or whether it is
46 practical to achieve further radionuclide removal. In conjunction with the review described in

1 Section 3.3, the reviewer should support a decision regarding the practicality of additional
2 radionuclide removal by following these review procedures:
3
4 • Confirm that the cost-benefit analysis includes a site-specific description of the affected
5 environment, alternative waste management strategies considered, and different
6 removal technologies evaluated. Confirm that estimated benefits of the different
7 radionuclide removal technologies and waste management strategies are clearly stated
8 (e.g., by estimating the reduction in risk or dose to the public, workers, and/or the
9 environment over time that is associated with various options).
10
11 • Determine whether information about radionuclide removal technologies used in the
12 cost-benefit analysis is reasonably current. Specifically, confirm that the most
13 appropriate technologies have been considered in the cost-benefit analysis (see
14 Section 3.3), and verify that the extent of radionuclide removal that could be achieved is
15 not underestimated.
16
17 • Verify that the potential benefits of additional removal are consistent with the results of
18 the performance assessment (Section 4) and inadvertent intruder analysis (Section 5).
19 Ensure that potential uncertainties in dose predictions are adequately represented in the
20 cost-benefit analysis (i.e., if doses could be significantly greater than doses assumed in
21 the cost-benefit analysis, the cost per averted dose of removing additional radionuclides
22 could be significantly lower than calculated in the cost-benefit analysis).
23
24 • If a detailed cost-benefit analysis is presented, ensure that the methods and values used
25 (e.g., discounting methods and discount rates) are appropriate. A reviewer may
26 compare the methods and values used with NRC guidance such as NUREG–1757
27 (NRC, 2003b, Vol. 2, Appendix N), if appropriate. Appropriate values for use in waste
28 determinations may be different from the values recommended in guidance applicable to
29 other types of sites (e.g., the period of performance).
30
31 • Determine whether the applicability of costs and benefits have been accounted for
32 appropriately. For example, if a technology must be developed to remove a component
33 from a particular waste stream (e.g., the removal of zeolite from a tank, or the removal of
34 technetium from salt waste), it may not be appropriate to attribute the entire cost of
35 technology development only to the proposed waste management activity if other
36 relevant benefits of the new technology (e.g., removal of zeolite from other tanks or
37 removal of technetium from other salt waste streams) are not included in the analysis.
38
39 • If a detailed cost-benefit analysis is presented, determine whether the analysis includes
40 relevant life-cycle costs, including variable costs (e.g., labor) and fixed costs (capital
41 equipment). Verify that estimated costs and benefits are expressed in monetary terms
42 where possible and expressed in constant dollars from the most recent year for which
43 adjustment data are available.
44
45 • If a detailed cost-benefit analysis is presented and if life-cycle costs are to be distributed
46 over time, verify that the cost-benefit analysis has used adequate discounting methods
47 such as those described in Office of Management and Budget (Office of Management
48 and Budget 2003, Circular A–4;1992, Circular A–94).
49

- If a detailed cost-benefit analysis is presented, determine whether the parameters used to quantify costs and benefits (e.g., transportation risks) are consistent with parameters presented in other NRC or OMB guidance [e.g., Table N.2 of NRC's guidance for License Termination (NRC, 2003b)]. If parameter values are not consistent with values provided in other NRC or OMB guidance and the parameters significantly affect the cost-benefit analysis results, determine whether an adequate basis has been provided to justify the use of alternate values.

- Verify that the timeframe(s) considered in the cost-benefit analysis is appropriate to the alternative waste management strategies and different removal technologies considered and include both short-term and long-term impacts.

- If DOE cites schedule impacts as a cost of additional radionuclide removal, determine the effect of the schedule impacts on the estimated cost or predicted doses to individuals (workers or members of the public).

- Evaluate whether additional protective measures could be taken to reduce worker exposure if radiological risks to workers are cited as a reason why additional radionuclide removal is impractical. If possible, determine whether taking additional protective measures to reduce worker exposure would contribute significantly to the cost of the proposed radionuclide removal activities.

- To the extent practical, compare the predicted costs and benefits of additional radionuclide removal to the range of costs per averted dose DOE has incurred to perform similar removal activities (e.g., the costs of previous efforts to remove radionuclides from the same tank or the costs to remove a similar type of waste from another tank).

3.5   Concentration Limits

Waste classification is a process used to ensure that waste is properly managed (e.g., LLW disposal in the near surface, HLW disposal in a geologic repository) to protect public health and safety. As discussed in Section 2.4.4, the NDAA, DOE Order 435.1 (DOE, 2001), and the West Valley Policy Statement (NRC, 2002) have different requirements regarding the classification of incidental waste according to the waste classifications described in 10 CFR 61.55. Essentially, the NDAA requires waste classification according to NRC's classification system, DOE Order 435.1 requires classification according to NRC's classification system or an alternate DOE-approved system, and the West Valley Policy Statement does not address waste classification with respect to incidental waste. Refer to Section 2.3 for the West Valley WIR criteria.

The risk from the near-surface disposal of radioactive waste is not just a function of concentration, but also the volume of waste and its accessibility. The potentially unique characteristics of incidental waste (e.g., depth of the waste, accessibility to intruders, radiological characteristics, spatial distribution) have been considered in the development of this guidance.

3.5.1 Areas of Review

To evaluate compliance with the applicable criteria of the NDAA and DOE Order 435.1, reviewers should assess DOE's classification of incidental waste according to the provisions of 10 CFR 61.55. Because NRC's waste classification system is based on radionuclide concentrations, radionuclide concentrations should be reviewed as described in Section 3.1. The concentration averaging provision of 10 CFR 61.55(a)(8) is applicable to the determination of the class of waste that has been mixed with or encapsulated within stabilizing material (e.g., residual tank waste stabilized with grout) or has otherwise been disposed of. Section 3.5.1.1 guides reviewers evaluating the concentration averaging provision of 10 CFR 61.55(a)(8) as it applies to incidental waste classification. Section 3.5.1.2 addresses the consultation requirements of the NDAA that are related to waste classification. The use of stabilizing material to protect inadvertent intruders is discussed in Section 5, and the use of stabilizing material to enhance site stability is discussed in Section 7.

3.5.1.1 Concentration Averaging

The guidance for concentration averaging in this document does not replace the guidance contained in the branch technical position (BTP) on concentration averaging and encapsulation (NRC, 1995b) for waste classification for the commercial disposal of LLW. The guidance is not intended to address all unique situations at DOE sites. However, this guidance is generally applicable to the following scenarios:

- Underground waste storage tanks including waste heels, cooling coils, and residual waste adhering to walls and other surfaces,

- Infrastructure used to support underground waste storage tanks such as transfer lines, transfer pumps, and diversion boxes,

- Waste removed from tanks that is processed or treated for disposal in a near surface disposal facility, and

- Other scenarios relating to waste determinations proposed by the DOE and accepted by the NRC.

Although the concentration averaging BTP was not written to address residual contamination of underground or buried structures or systems, the fundamental principles of waste classification contained within the BTP apply to these systems. In addition, the fundamental principles supporting the development of the waste classification system of 10 CFR Part 61 also apply to waste determinations. This concentration averaging guidance clarifies the fundamental principles presented in the BTP and provides specific examples that may be pertinent to DOE waste determinations. The acceptable concentration averaging methods to classify waste for waste determinations are based on the following fundamental principles introduced in the BTP and in the analysis supporting the development of Part 61:

- Extreme quantities of uncontaminated materials should not be averaged with residual contamination solely for the purpose of waste classification.

1 •   Mixtures of residual waste and materials can use a volume- or mass-based average
2   concentration if it can be demonstrated that the mixture is reasonably well-mixed. The
3   type of averaging (volume- or mass-based) should be consistent with the units of
4   concentration specified for each radionuclide in Tables 1 and 2 of 10 CFR 61.55.
5
6 •   Credit can be taken for stabilizing materials added for the purpose of immobilizing the
7   waste (not for stabilizing the contaminated structure) even if it cannot be demonstrated
8   that the waste and stabilizing materials are reasonably well mixed, when the
9   radionuclide concentrations are likely to approach uniformity in the context of applicable
10   intruder scenarios.
11
12 •   Waste classification can use a risk-informed approach that calculates waste
13   classification based on the unique characteristics of a specific site (e.g., depth to waste).
14   The intruder scenarios used for waste classification should be conservative and
15   consistent with those discussed in Section 3.5 of this guidance document.
16
17 •   Other provisions for the classification of residual waste may be acceptable if, after
18   evaluation of the specific characteristics of the waste, disposal site and method of
19   disposal, the performance objectives in 10 CFR Part 61, Subpart C can be demonstrated
20   with reasonable assurance.
21
22 •   Regardless of the averaging that is performed for waste classification purposes, the
23   performance assessment or other approach used to demonstrate compliance with the
24   performance objectives of 10 CFR Part 61, Subpart C must consider the actual
25   distribution of residual contamination in the system when estimating release rates to the
26   environment and exposure rates to onsite (inadvertent intruders) and offsite public
27   receptors. Conservative assumptions regarding the distribution of contamination
28   are appropriate.
29
30 The purpose of these principles is to prevent arbitrary or incorrect classification of materials that
31 may result in near-surface disposal of materials that are not suitable for near-surface disposal.
32 The potentially unique characteristics of incidental waste (e.g., depth of the waste, accessibility
33 to intruders, radiological characteristics, spatial distribution) have been considered in the
34 development of this guidance. Appropriate concentration averaging may indicate that waste
35 exceeds Class C concentration limits. Waste that exceeds Class C concentration limits may be
36 suitable for near-surface disposal, but the evaluation of the suitability must involve independent
37 analysis such as would be performed by the NRC under 10 CFR 61.58. NRC staff would
38 evaluate the safety of the near-surface disposal of waste that exceeds Class C concentration
39 limits on a case-by-case basis. Waste concentration is, in some cases, only one of many
40 factors that can influence risk. Waste that is greater than Class C may be determined to be
41 incidental waste and may be safely managed with near-surface disposal if it can be
42 demonstrated that the performance objectives of 10 CFR Part 61, Subpart C, are satisfied.
43 Conversely, waste may be determined to be less than Class C; however, the waste may be
44 inappropriate for near-surface disposal because the performance objective of 10 CFR 61.41
45 cannot be satisfied. The methods described in the three categories that follow can be used to
46 determine the waste classification of waste residuals.
47
48 This guidance presents three categories of calculations (Categories 1, 2, and 3) to determine
49 waste classification for the concentrations of radionuclides in waste. The first pertains to cases

1 in which the waste can be mixed and is fairly homogeneous. The second pertains to cases in
2 which the waste cannot be well mixed and is stabilized in place to satisfy the requirements of
3 10 CFR 61.56. The third category is a risk-informed approach developed for incidental waste to
4 provide flexibility in waste classification calculations, recognizing the differences between the
5 disposal of incidental waste and commercial LLW and recognizing site-specific conditions.
6 Category 1 is the simplest approach, while Categories 2 and 3 account for more unique site
7 information but also require more effort to apply.
8
9 As the first principle indicates, extreme measures should not be taken when performing
10 concentration averaging to determine waste classification. The extremity of a measure is
11 relative to the particular averaging approach applied (i.e., Category 1, 2, or 3). The averaging
12 approaches are discussed in this guidance. For Category 1, extreme measures may include but
13 are not limited to (1) averaging assumptions that are inconsistent with the physical distribution
14 of radionuclides over the averaging volume or mass and (2) deliberate blending of lower
15 concentration waste streams with high activity waste streams solely to achieve waste
16 classification objectives, although blending may be needed for waste management purposes.
17 For Category 2, extreme measures may include but are not limited to (1) averaging over
18 stabilizing material volume or masses that are not needed to stabilize the waste per the
19 10 CFR 61.56 stability requirements and (2) averaging over stabilizing material volume or
20 masses that are not homogeneous from the context of the intruder scenarios.
21
22 Other provisions may be used for representing the distribution of radionuclides in waste and the
23 disposal system in the performance assessment calculations to determine the suitability of
24 near-surface disposal according to 10 CFR 61.58, but these other provisions do not pertain to
25 the determination of whether a waste is Class A, Class B, Class C, or greater than Class C as
26 defined in 10 CFR 61.55.
27
28 Category 1—Physical Homogeneity
29
30 In general, waste will have been processed to the maximum extent practical and will have been
31 stabilized so that there is reasonable assurance that the performance objectives of
32 10 CFR Part 61, Subpart C, can be achieved. The concentrations of radionuclides in the waste
33 used to estimate waste classification can be based on the average concentration calculated
34 from the total volume or mass of the waste and processing or stabilizing materials if the
35 materials are reasonably well mixed. The type of averaging (volume- or mass-based) should be
36 consistent with the units of concentration specified for each radionuclide in Tables 1 and 2 of
37 10 CFR 61.55. The weight or volume of the container should not be included in the calculation
38 of average concentrations. The primary consideration for application of a Category 1 approach
39 is whether the distribution of radionuclides within the final wasteform is reasonably
40 homogeneous. Technical basis should be provided (e.g., sampling results, engineering
41 experience, operational constraints) to demonstrate that the waste is reasonably well mixed.
42 The preferred method to demonstrate homogeneity would be to provide a statistical measure
43 (e.g., coefficient of variation) of the variability of concentration within the waste, although this
44 may not always be practical. For homogeneous mixtures, the classification of waste residuals
45 may be based on total volume or mass of the final wasteform. If additional averaging (e.g., as in
46 the examples in Category 2) is not applied, waste with radionuclide concentrations after mixing
47 that are greater than the values provided in Tables 1 and 2 of 10 CFR 61.55 would be
48 considered to be greater than Class C waste.
49

1 Mixing within waste or of waste with stabilizing materials may be needed for a variety of
2 reasons. Mixing of waste and stabilizing materials may help to reduce release rates to achieve
3 performance objectives. With respect to the principles of the BTP, mixing with excessive
4 amounts of stabilizing materials solely to reduce the waste concentrations to alter waste
5 classification should not be performed. In most cases, the ratio of the unstabilized to stabilized
6 radionuclide concentrations would not be significantly greater than a factor of 10 for waste
7 classification purposes. For unstabilized waste that cannot be selectively treated or removed,
8 mixing (within waste, not between waste streams) to facilitate homogenization of radionuclide
9 concentrations is appropriate. For example, mixing may be used to reduce the variability in
10 concentrations within a layer of tank waste that cannot be removed for further treatment.
11
12 Example 1-1
13
14 *Liquid waste is removed from a tank, and additional fluids are added to adjust the chemistry for*
15 *processing. Cement and fly ash are mixed with the resultant liquid in an industrial mixer to form*
16 *a grout that is placed in disposal containers. The concentration of radionuclides for determining*
17 *waste classification is based on the total volume or mass of the final wasteform.*
18
19 Example 1-2
20
21 *Reducing grout is added to stabilize a tank heel (the residual layer of waste in the bottom of a*
22 *tank). The waste residuals in the tank are flocculated solids suspended in a liquid phase that*
23 *can be mobilized with the tank transfer equipment. However, the solids cannot be removed with*
24 *the existing equipment. The reducing grout has a relatively high viscosity such that the*
25 *flocculated solid residuals and remaining waste liquids can be mixed with the grout using the*
26 *transfer equipment prior to setting of the grout. The concentration of radionuclides for waste*
27 *classification is based on the total volume or mass of the waste and the reducing grout in which*
28 *the waste is mixed. Additional reducing grout into which little or no waste is mixed should not*
29 *be included in the total mass or volume used for concentration averaging.*
30
31 Category 2—Stabilization to Satisfy 10 CFR 61.56
32
33 Stabilization can limit exposure to an inadvertent intruder because it provides a recognizable
34 and non-dispersible waste. For solidified liquids and solids, Section 3.2 of the BTP provides for
35 the concentration of the radionuclides to be determined based on the volume or weight of the
36 solidified mass, which is defined here to be the amount of material needed to stabilize the
37 liquids or dispersible solids to satisfy 10 CFR 61.56. Liquid waste must be solidified or
38 packaged in sufficient absorbent material to absorb twice the volume of the liquid
39 (10 CFR 61.56). However, the stabilizing material is not to be interpreted as bulk material
40 added to fill void space in the storage system or to fill void space in the equipment containing
41 the waste. Stabilization is determined with respect to the waste and not the entire disposal
42 system or unit. While stabilization of the entire disposal unit (e.g., a tank) may be necessary to
43 meet the performance objectives, it generally would not be needed to make the residual waste
44 recognizable and nondispersible.
45
46 Waste concentrations are calculated based on the volume or mass of material needed to be
47 added to liquids or dispersible solids to solidify or encapsulate them. The concentration of the
48 stabilized waste (waste plus stabilizing material) should generally be within a factor of 10 of the
49 concentration of the unstabilized waste on either a mass or volume basis. The factor of 10 is

1 based on the amount of material typically needed to stabilize waste in a solidified wasteform,
2 consistent with the principle that concentration averaging is appropriate when wastes require
3 stabilization. Most commonly envisioned stabilization techniques use cementitious materials,
4 and most cementitious wasteforms can readily achieve a minimum of 10 mass percent waste
5 loading. Additional stabilizing materials may be needed for the system or structures however,
6 they would not generally be needed for waste stabilization. Therefore, it would not be
7 consistent with the principles of classification for this category to include those materials in the
8 concentration averaging calculations. A more risk-informed approach that considers site-
9 specific conditions is found under Category 3.
10
11 For thin layers of contamination on surfaces, especially vertical surfaces, the average
12 concentration may be based on the volume or mass of a layer of stabilizing material that would
13 be needed to stabilize the waste, as discussed previously. This does not mean that averaging
14 can be performed over all materials added to fill void space in the structure. This approach is
15 justified because the concentrations would be expected to approach homogeneity with respect
16 to the intruder scenarios, and the main justification for the classification system is to protect the
17 inadvertent intruder. The concentration values found in Tables 1 and 2 of 10 CFR 61.55 were
18 derived assuming the total volume of waste exhumed by the intruder is at those concentrations
19 (although dilution of waste was considered after the waste was exhumed) therefore, a thin layer
20 of more concentrated material averaged over the same exhumed volume would achieve a
21 similar level of protection. Specific averaging volumes are not provided for this category
22 because of the site-specific nature of the waste and site-specific considerations for intruder
23 scenarios. The type of averaging (volume- or mass-based) should be consistent with the units
24 of concentration specified for each radionuclide in Tables 1 and 2 of 10 CFR 61.55. If a
25 Category 2 approach is not flexible enough for a particular waste determination, then a
26 Category 3 approach should be considered.
27
28 Example 2-1
29
30   A tank contains a heel that is 2.5 cm [1 in] thick and is composed of liquids and dispersible
31   solids. A 20-cm [8-in]-thick layer of reducing grout is needed to stabilize the waste, and an
32   additional 300 cm [120 in] of high-strength grout is added to fill void space and to provide an
33   intruder barrier. The concentration of radionuclides would be calculated by averaging over the
34   20-cm [8-in]-thick layer of reducing grout. Use of a 20-cm [8-in] layer of reducing grout in the
35   concentration calculation is based on the amount of grout that would be needed to stabilize the
36   waste if it could be removed from the tank and made into a stable wasteform. The
37   concentration of the stabilized waste (waste plus stabilizing material) would generally be within
38   a factor of 10 of the concentration in the unstabilized waste on either a mass or volume basis.
39
40 Example 2-2
41
42   The walls of a waste storage tank have a thin layer {0.1 cm [0.04 in]} of residual contamination
43   that is not easily removed. The tank walls are 1 cm [0.4 in] thick, and the tank is contained
44   within a 0.5-m [1.6-ft]-thick vault. The contamination is not easily dispersed into the
45   environment and is located underground. Closure of the storage tank will involve filling the tank
46   and all void space with grout. The concentration of the waste for waste classification is
47   calculated based on the thickness of the contamination and a 1-cm [0.4-in]-thick layer of
48   stabilizing grout. Use of a 1-cm [0.4-in] layer of grout in the concentration calculation is based
49   on the assumption that formation of a stable wasteform is accomplished by incorporating the

1 *0.1-cm [0.04-in] layer of residual waste into a cementitious wasteform at a mass loading of*
2 *approximately 10 percent. The concentrations of the thin layer would be reduced by a factor of*
3 *10 for estimating waste classification if a volume basis were used.*
4
5 Category 3—Site-Specific Averaging
6
7 This approach, accounting for site-specific conditions, may be used to determine whether
8 incidental waste is greater than Class C. A description of the calculations used to develop the
9 example generic averaging expressions found in this section is provided in Appendix B.
10
11 The risk from the near-surface disposal of radioactive waste is not just a function of
12 concentration, but also volume and accessibility. The Category 3 approach to concentration
13 averaging attempts to account for the volume of waste, concentration of waste, and accessibility
14 of the waste. Note that a number of factors (including those listed previously) specific to
15 commercial LLW disposal were considered in the Part 61 analysis to develop waste
16 classification concentration limits (NRC, 1981). Some of those factors may not apply to a
17 specific incidental waste site, or more recent information may suggest taking a different
18 approach. It is not as simple as applying a new dilution factor for an incidental waste scenario
19 and modifying the 10 CFR 61.55 waste classification concentration limits directly.
20
21 Although accessibility of the waste (e.g., depth and intruder barriers) is considered for
22 concentration averaging applied to incidental waste, managing of intruder risk with complex,
23 engineered facilities to prohibit access of the waste for very long periods of time (more than
24 500 years) is not likely to be practical. Thus, there are practical limits as to what types of waste
25 are suitable for near-surface disposal. Waste classification is a process used to ensure that
26 waste is properly managed (e.g., LLW disposal in the near surface, HLW disposal in a geologic
27 repository) to protect public health and safety.
28
29 Although multiple scenarios were considered, the limiting intruder scenario that was used (in
30 deriving the concentration limits for waste classification found in Table 1 and Table 2 of
31 10 CFR 61.55) was an intruder construction scenario (NRC, 1981). This scenario involved the
32 excavation of a foundation for a house. Approximately 232 m^3 [8,190 ft^3] of waste was
33 assumed to be exhumed, and the excavation was assumed to be to a depth of 3 m [10 ft]. This
34 scenario was appropriate for the commercial disposal of LLW, which at the time was envisioned
35 to be in shallow trenches covering a large area (e.g., thousands of square meters). The
36 commercial disposal of Class C waste is required to have the top of the waste a minimum of
37 5 m [16 ft] below the top surface of the cover, or it must be disposed of with intruder barriers that
38 are designed to protect against an inadvertent intrusion into the waste for at least 500 years.
39 Tables 1 and 2 were developed so that waste generators and disposers could safely manage
40 and dispose of waste of various concentrations and compositions. The Table 1 and 2 values
41 were developed making certain assumptions regarding the exposure scenarios and radiological
42 distribution of waste that a commercial LLW facility would likely receive, and the analysis used
43 the most current information on dosimetry and disposal technology (NRC, 1981). The
44 underlying assumption of the values in Tables 1 and 2 of 10 CFR 61.55 is that the concentration
45 limits and disposal requirements ensure that an inadvertent intruder would not receive a dose
46 exceeding an equivalent of 5 mSv [500 mrem] to the whole body. For example, an intruder
47 analysis assuming an excavation scenario and Class C waste disposed of at a depth less than
48 5 m [16 ft] without an intruder barrier results in a dose substantially in excess of 5 mSv
49 [500 mrem].

1
2   Incidental waste disposal may be similar to commercial disposal of LLW in some aspects and
3   may be substantially different in other aspects. A Category 3 approach allows a reviewer to
4   account for these differences. Figure 3-1 summarizes the 10 CFR Part 61 analysis for the
5   excavation scenario and a hypothetical disposal configuration for incidental waste in a tank.
6   Whereas the excavation scenario is appropriate for relatively shallow waste, the excavation
7   scenario is not appropriate for deeper waste that is significantly below the excavation depth. In
8   these cases, a drilling scenario may be more appropriate.
9
10   The relationship between waste concentration and disposal depth was recognized in the Draft
11   Environmental Impact Statement for 10 CFR Part 61 and updates to the impacts analysis (NRC,
12   1981, 1986). Consideration was given to excavations that may be deeper than 5 meters, such
13   as for an industrial building. A subjective probability of 10 percent was assigned to the
14   likelihood that a deeper excavation would be inadvertently constructed at a LLW facility resulting
15   in disruption of the waste. Consideration was also given to robust intruder barriers and the
16   impacts they may have on safety. As noted above, robust intruder barriers are required for the
17   shallow commercial disposal of Class C waste. A robust intruder barrier was assumed to delay
18   an intrusion event for up to 500 years. However, technological limitations and uncertainties of
19   barrier performance were recognized such that even in the case of a "hot waste facility" (an
20   engineered facility that would be constructed of high-strength reinforced concrete to contain
21   high activity waste such as sealed sources), credit for the concrete as an intruder barrier for
22   time periods exceeding 1,000 years was not believed to be reasonable. For near-surface
23   disposal there were considered to be technological limitations to preventing intrusion for very
24   long periods of time.
25
26   The calculations to develop the Table 1 and Table 2 concentration limits in 10 CFR 61.55 to
27   define Class A, B, and C waste were based on a number of important assumptions and
28   considerations. Although a requirement is provided that Class C waste must either be
29   commercially disposed of at depths greater than 5 m [16 ft] or have a robust intruder barrier
30   that lasts for 500 years, the calculations assumed an excavation scenario for the intruder. The
31   calculations also assumed that a large volume fraction of the waste at a commercial LLW
32   facility would not be Class C waste. Therefore, credit for dilution or mixing of different
33   concentrations of waste was provided. The current internal dosimetry at the time, Report 2 of
34   the International Commission on Radiological Protection (ICRP–2) (ICRP, 1959), was used in
35   deterministic analyses.
36
37   Incidental waste may include many different types of waste in a variety of disposal
38   configurations. For example, incidental waste may include residual waste in piping that is near
39   the land surface and does not have a robust intruder barrier. Residual waste in a tank that is
40   much deeper than 5 m [16 ft] from the land surface and does have a robust intruder barrier may
41   also be considered to be incidental waste. Guidance for waste classification for incidental
42   waste needs to be flexible to handle these different aspects.
43
44   The conceptual approach and example Category 3 concentration averaging expressions were
45   developed using a risk-informed approach to account for (1) depth of waste, (2) intruder
46   barriers, (3) current dosimetry, and (4) propagation of uncertainty into the concentration limits.
47   The example averaging expressions represent a conversion from the analyses used to support
48   the development of Tables 1 and 2 of 10 CFR 61.55 [Figure 3-1(a)] and calculations more
49   appropriate for incidental waste [Figure 3-1(b)]. Different approaches can be applied

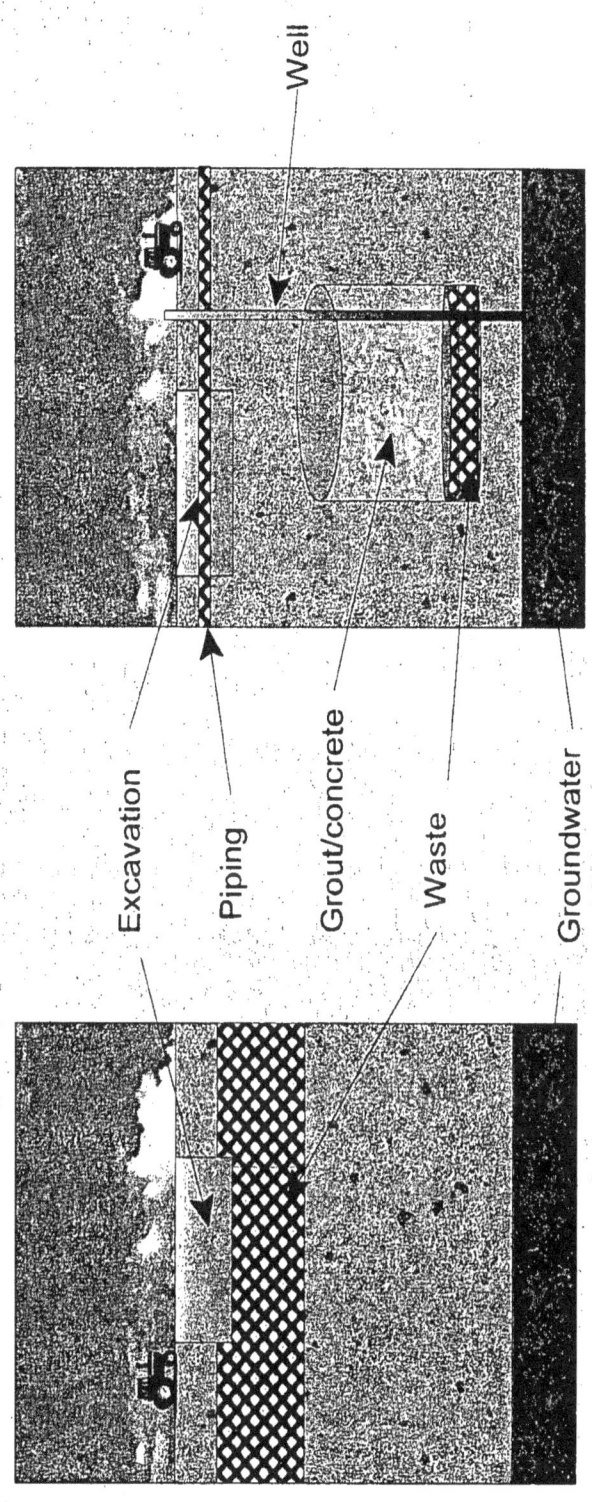

Incidental Waste Intruder Scenario for Tank Residuals or Ancillary Equpiment

Part 61 Intruder Construction Scenario

Well

Excavation

Piping

Grout/concrete

Waste

Groundwater

Probabilistic or deterministic calculations
Dosimetry - ICRP 26 and 30
Site-specific parameter values or distributions
(b)

Deterministic calculations
Dosimetry - ICRP 2
Generic parameter sets
(a)

Figure 3-1. Comparison of Key Assumptions of (a) the Intruder Excavation Scenario for Part 61 and (b) Intruder Scenarios for Tank Residual Waste

2

3

4

1　depending on the depth of the waste and whether a robust intruder barrier has been installed.
2　Four different example averaging expressions have been developed to provide flexible
3　benchmarks for the staff to use when reviewing Category 3 concentration averaging
4　approaches. It is expected that DOE will develop its own calculations to support a
5　Category 3 approach.
6
7　Depth to waste and the use of a robust intruder barrier are two primary conditions that
8　determine the type of intruder scenario most appropriate to estimate waste classification. The
9　burial depth condition is evaluated considering cover thicknesses projected over the analysis
10　period (e.g., processes that may decrease the cover thickness over time should be considered).
11　A *robust* intruder barrier is defined as one that will prevent intrusion into the waste for
12　500 years. Regardless of the approach that applies, the waste should be immobilized in a solid
13　physical form such that it would be recognizable as being non-native materials and it would not
14　be easily dispersed into the environment (consistent with 10 CFR 61.56). The sum of fractions
15　approach, consistent with the description provided in 10 CFR 61.55, should be applied for waste
16　containing multiple radionuclides.
17
18　The review of site-specific averaging for waste classification should apply to the following
19　conditions and assumptions:
20
21　•　If the waste is buried at a minimum depth of less than 5 m [16 ft] and a robust intruder
22　　barrier is not installed, then an excavation scenario should be assumed and applied at
23　　the time of maximum dose (typically 100 years) following closure of the system. Both
24　　acute (e.g., the receptor who builds the residence) and chronic (e.g., the receptor who
25　　lives in the residence after construction) exposure of receptors should be evaluated and
26　　the more limiting receptor type should be used to define waste classification.
27
28　•　If the waste is buried at a minimum depth of less than 5 m [16 ft] and a robust intruder
29　　barrier is installed, then an excavation scenario should be assumed and applied at the
30　　time of maximum dose (typically 500 years) following closure of the system. A robust
31　　intruder barrier is defined as one that will last 500 years. Both acute (e.g., the receptor
32　　who builds the residence) and chronic (e.g., the receptor who lives in the residence after
33　　construction) exposure of receptors should be evaluated, and the more limiting receptor
34　　should be used to define waste classification.
35
36　•　If the waste is buried at a minimum depth of more than 5 m [16 ft] and a robust intruder
37　　barrier is not installed, then a drilling scenario should be assumed and applied at the
38　　time of maximum dose (typically 100 years) following closure of the system. Both acute
39　　(e.g., the receptor who constructs the well) and chronic (e.g., the receptor who lives in a
40　　residence located where the cuttings or material from construction of the well are
41　　dispositioned) exposure of receptors should be evaluated, and the more limiting type
42　　should be used to define waste classification.
43
44　•　If the waste is buried at a minimum depth of more than 5 m [16 ft] and a robust intruder
45　　barrier is installed, then a drilling scenario should be assumed and applied at the time of
46　　maximum dose (typically 500 years) following closure of the system. Both acute
47　　(e.g., the receptor who constructs the well) and chronic (e.g., the receptor who lives in a
48　　residence located where the cuttings or material from construction of the well are

dispositioned) exposures of receptors should be evaluated, and the more limiting type should be used to define waste classification.

- If a scenario other than those listed is more limiting and is reasonably likely to occur after closure, then it should be used to determine waste classification with a site-specific averaging approach.

Staff should verify that DOE applied these assumptions when performing site-specific concentration averaging. Pathways considered and parameters used for each scenario should be provided. Staff should verify that scenarios selected are consistent with Table 3-2. For waste classification purposes, one of the four scenarios from Table 3-2 or a more limiting scenario, must be selected if a Category 3 approach is used. The scenarios in Table 3-2 are believed to be the limiting scenarios out of the possible combinations, as waste access time and waste disruption process are dictated by the depth of waste and the presence or absence of a robust intruder barrier. The waste classification system is, in part, designed to establish a protective upper limit to the concentration (and quantity) of material that may be suitable for near-surface disposal. Quantity of waste can be either explicitly represented in the analysis or implicitly represented with assumptions about the disposal facility type and configuration. Use of reasonably conservative scenarios for waste classification purposes ensures that radioactive waste is described in a proper risk context, radioactive waste is disposed of with appropriate controls, and classification calculations are reviewed with an appropriate level of effort to ensure public health and safety will be protected. Because the Category 3 approach to concentration

Table 3-2. Assumed Conditions for the Four Scenarios Used to Develop Averaging Expressions			
Scenario	**Typical Waste Access Time* (yr)**	**Waste Disruption Process**	**Receptor Type**
Shallow waste, no intruder barrier	100	Residential Construction	Construction worker—acute or Resident—chronic
Shallow waste, intruder barrier	500	Residential Construction	Construction worker—acute or Resident—chronic
Deep waste, no intruder barrier	100	Well Drilling	Well driller—acute or Resident—chronic
Deep waste, intruder barrier	500	Well Drilling	Well driller—acute or Resident—chronic
*The maximum dose may not occur at the earliest possible waste access time, depending on the decay of parent and in-growth of daughter radionuclides. For fission products, peak dose will likely occur at the earliest waste access time for a given scenario.			

1 averaging and thus waste classification involves a site-specific intruder analysis, it is unlikely
2 that waste estimated to be greater than Class C waste using a Category 3 approach would
3 satisfy the 10 CFR 61.42 performance objective.
4
5 Note that a number of factors specific to commercial LLW disposal were considered in the
6 10 CFR Part 61 analysis to develop waste classification concentration limits (NRC, 1981).
7 Some of those factors may not apply to a specific incidental waste site, or more recent
8 information may suggest a different approach should be taken. Adapting the waste
9 classification approach considered during the development of 10 CFR Part 61 to incidental
10 waste is not as simple as applying a new dilution factor for an incidental waste scenario and
11 modifying the 10 CFR 61.55 waste classification concentration limits directly.
12
13 Example averaging expressions were developed to provide a benchmark for staff to use to
14 review site-specific averaging for waste classification under a variety of scenarios. The example
15 averaging expressions (developed assuming a generic site and receptor characteristics) were
16 developed with moderately conservative assumptions to ensure adequate protection under most
17 conditions. For example, as explained in Appendix B, the limiting short-lived and the limiting
18 long-lived radionuclides were used to develop the constants in the example averaging
19 expressions that then would be applied to all radionuclides of that type (i.e., short-lived or long-
20 lived). However, site-specific conditions, particularly for water-dependent pathways, can affect
21 the amount of averaging that may be appropriate at a site. Site-specific variability can result in
22 an order of magnitude or more range for the constants in the example averaging expressions.
23 *Therefore, even if the resultant site-specific averaging applied for incidental waste is consistent*
24 *with that estimated by staff with the relevant example averaging expression found below, the*
25 *averaging analysis should be carefully reviewed. The averaging expressions (benchmarks) are*
26 *not to be used as the basis for site-specific averaging for waste classification. They are for staff*
27 *use to determine when site-specific concentration averaging calculations DOE performed may*
28 *require additional staff review effort.* A description of the development of the benchmarks are
29 found in Appendix B.
30
31 **If waste is buried at depths greater than 5 m [16 ft] and a robust intruder barrier**
32 **is installed:**
33
34 For Table 1 or Table 2 radionuclides, individual radionuclide contribution to the sum of fractions
35 may be estimated with the following equation (drilling is assumed to occur at 500 years)
36

$$RC_i = \left(\frac{WC_i}{Table\_Value_i} \right) * \left(\frac{Waste\_thickness}{Drill\_depth} \right) * 7 \qquad (3.5\text{-}1)$$

38 where
39
40 i   =   radionuclide index
41 RC_i =   radionuclide classification factor (unitless)
42 WC_i =   concentration in the waste for radionuclide i in units consistent
43   with the appropriate Table value in 10 CFR 61.55
44 $Table\_Value_i$ =   Class C concentration limit in the appropriate Table of
45   10 CFR 61.55 for radionuclide i
46 Waste_thickness   =   thickness of the waste (m)

1 Drill_depth   =   depth an intruder would likely install a well to in order to recover
2   resources (m)
3
4 RC$_i$ represents the contribution to the Table 1 or Table 2 sum of fractions [see
5 10 CFR 61.55 (a)(7)] for each radionuclide. The derivation of the constant, 7, in the
6 expression is discussed in Appendix B. In most cases, the resource being recovered would
7 be water; however, exploitation of other natural resources should be considered. The
8 long-lived radionuclides are only moderately more limiting than the short-lived radionuclides for
9 this example, therefore the same averaging can be applied to each group of radionuclides.
10
11 **If waste is buried at depths greater than 5 m [16 ft] and a robust intruder barrier is**
12 **not installed:**
13
14 For Table 1 radionuclides, individual radionuclide contribution to the sum of fractions may be
15 estimated with the following equation (drilling is assumed to occur at 100 years)

$$16 \qquad RC_i = \left(\frac{WC_i}{Table1_i} \right) * \left(\frac{Waste\_thickness}{Drill\_depth} \right) * 7 \qquad\qquad (3.5\text{-}2)$$

17 where
18
19 i   =   radionuclide index
20 RC$_i$ =   radionuclide classification factor (unitless)
21 WC$_i$ =   concentration in the waste for radionuclide i (in Ci/m^3 or nCi/g)
22 Table1$_i$ =   concentrations in Table 1 of 10 CFR 61.55 for radionuclide i (in
23   Ci/m^3 or nCi/g).
24 Waste_thickness   =   thickness of the waste (m)
25 Drill_depth   =   depth an intruder would likely install a well to in order to recover
26   resources (m)
27
28 The RC$_i$ represents the contribution to the Table 1 sum of fractions [see 10 CFR 61.55 (a)(7)]
29 for each radionuclide. The derivation of the constant, 7, in the expression is discussed in
30 Appendix B.
31
32 For Table 2 radionuclides, individual radionuclide contribution to the sum of fractions may be
33 estimated with the following equation

$$34 \qquad RC_i = \left(\frac{WC_i}{Table2\_Column3_i} \right) * \left(\frac{Waste\_thickness}{Drill\_depth} \right) * 17{,}000 \qquad\qquad (3.5\text{-}3)$$

35 where
36
37 i   =   radionuclide index
38 RC$_i$ =   radionuclide classification factor (unitless)
39 WC$_i$ =   concentration in the waste for radionuclide i (Ci/m^3)
40 Table2_Column3$_i$ =   concentrations in Column 3 of Table 2 of 10 CFR 61.55 for
41   radionuclide i (Ci/m^3)
42 Waste_thickness   =   thickness of the waste (m)
43 Drill_depth   =   depth an intruder would likely install a well to in order to recover
44   resources (m)

1 The RC_i represents the contribution to the Table 2 sum of fractions for each radionuclide. The
2 derivation of the constant, 17,000, in the expression is discussed in Appendix B.
3
4 **If waste is buried at <u>depths less than 5 m [16 ft]</u> and a robust intruder barrier <u>is</u> installed:**
5
6 For *all radionuclides,* individual radionuclide contribution to the sum of fractions may be
7 estimated with the following equation (excavation is assumed to occur at 500 years)

8 $$RC_i = \left(\frac{WC_i}{Table\_Value_i}\right) * \left(\frac{Waste\_volume}{3\,m^3}\right) \qquad (3.5\text{-}4)$$

9 where
10
11 i   =   radionuclide index
12 RC_i =   radionuclide classification factor (unitless)
13 WC_i =   concentration in the waste for radionuclide i
14 $Table\_Value_i$ =   Class C concentration limit in the appropriate table of
15   10 CFR 61.55 for radionuclide i
16 Waste_volume   =   volume of waste (in units of m^3) in an excavation volume of 600 m^3
17
18 The values for WC_i and $Table\_Value_i$ should be expressed in the same units. The RC_i
19 represents the contribution to the sum of fractions for each radionuclide. The derivation of the
20 constant, 3, in the expression above is discussed in Appendix B. When Waste_volume is
21 expressed in m^3, the result of expression Eq. (3.5-4) is unitless.
22
23 The waste classification should also be estimated using the approach for waste buried at
24 depths greater than 5 m [16 ft] with a robust intruder barrier installed, and the more limiting
25 value should be used.
26
27 **If waste is buried at depths <u>less than 5 m [16 ft]</u> and a robust intruder barrier <u>is</u>**
28 **<u>not</u> installed:**
29
30 For *Table 1 radionuclides,* individual radionuclide contribution to the sum of fractions may be
31 estimated with the following equation (excavation is assumed to occur at 100 years)

32 $$RC_i = \left(\frac{WC_i}{Table1\_Value_i}\right) * \left(\frac{Waste\_volume}{3\,m^3}\right) \qquad (3.5\text{-}5)$$

33 where
34
35 i   =   radionuclide index
36 RC_i =   radionuclide classification factor (unitless)
37 WC_i =   concentration in the waste for radionuclide i
38 $Table1\_Value_i$ =   Class C concentration limit in Table 1 of 10 CFR 61.55 for
39   radionuclide i
40 Waste_volume   =   volume of waste (in units of m^3) in an excavation volume of 600 m^3
41

1   The values for WC_i and Table1_Value$_i$ should be expressed in the same units. The RC_i
2   represents the contribution to the sum of fractions for each radionuclide. The derivation of the
3   constant, 3, in the expression above is discussed in Appendix B.
4
5   For **Table 2 radionuclides**, individual radionuclide contribution to the sum of fractions may be
6   estimated with the following equation

7
$$RC_i = \left(\frac{WC_i}{\textbf{Table2\_Value}_i} \right) * Waste\_volume * 1{,}000\, m^{-3} \qquad (3.5\text{-}6)$$

8   where
9
10   i   =   radionuclide index
11   RCi   =   radionuclide classification factor (unitless)
12   WC_i =   concentration in the waste for radionuclide i
13   Table2_Value$_i$ =   Class C concentration limit in Table 2 of Part 61 for radionuclide i
14   Waste_volume   =   volume of waste (in units of m³) in an excavation volume of 600 m³
15
16   The values for WC_i and Table2_Value$_i$ should be expressed in the same units. The RC_i
17   represents the contribution to the sum of fractions for each radionuclide. The derivation of the
18   constant, 1,000, in the expression above is discussed in Appendix B.
19
20   The waste classification should also be estimated using the approach for waste buried at depths
21   greater than 5 m [16 ft] without a robust intruder barrier installed, and the more limiting value
22   should be used.
23
24   <u>Example 3-1</u>
25
26   *A waste tank contains a 5-cm [2-in] layer of waste after cleaning, and the bottom of the tank is*
27   *roughly 10 m [33 ft] below the ground surface. Erosion processes are estimated to be negligible*
28   *for the foreseeable future. The average Np-237 concentration in the waste is 3,000 nCi/g. The*
29   *depth to the water table is 20 m (66 ft) from the ground surface at the tank location, and a*
30   *high-strength grout is emplaced in the tank to stabilize the residual waste and provide a robust*
31   *intruder barrier. Only a single radionuclide is used in this example to illustrate the calculation;*
32   *the waste in most real world problems would contain numerous radionuclides and would require*
33   *using a sum of fractions approach be used.*
34
35   If the waste classification approach for a commercial LLW facility was used, the residual waste
36   would be greater than Class C because the concentration exceeds 100 nCi/g (Table 1 of
37   10 CFR 61.55). However, DOE applied a Category 3 approach and provided detailed
38   documentation of the waste classification calculations. Waste classification was based on an
39   intruder driller scenario that was assumed to occur at 100 years after facility closure. The waste
40   access time of 100 years was conservatively selected to account for uncertainty in the ability of
41   the robust intruder barrier to limit contact with the waste. DOE estimated the waste class to be
42   less than Class C (sum of fractions = 0.81) in its calculations.
43
44

1　 _Using the appropriate Category 3 expression, the staff calculated the waste classification for_
2　 _this scenario to be less than Class C_

3

$$RC_{237\,Np} = \frac{3{,}000\,\dfrac{nCi}{g}}{100\,\dfrac{nCi}{g}} \times \frac{0.05m}{20m} \times 7 = 0.525 \qquad (3.5\text{-}7)$$

4　 _Because the benchmark result is reasonably consistent with the DOE estimate, staff need not_
5　 _perform independent waste classification calculations and can perform a standard review of the_
6　 _DOE document._
7
8　 Other Provisions
9
10　 10 CFR 61.58 allows the Commission to authorize other provisions to classify and characterize
11　 waste, if after evaluating the specific characteristics of the waste, disposal site, and method of
12　 disposal, it finds reasonable assurance of compliance with the performance objectives in
13　 Subpart C. Demonstration that the performance objectives can be satisfied would involve a
14　 site-specific analysis (e.g., performance assessment). 10 CFR 61.58 was intended to allow the
15　 NRC the flexibility of establishing alternate waste classification schemes when justified by
16　 site-specific conditions and does not affect the generic waste classifications established in
17　 10 CFR 61.55. Thus, if the results of concentration calculations performed in a manner
18　 consistent with the principles and examples described previously in this document indicate that
19　 radionuclide concentrations in the waste exceed Class C limits, then the waste is greater than
20　 Class C waste for waste classification purposes. If it can be demonstrated that the
21　 performance objectives of 10 CFR Part 61 can be satisfied, then the waste would be suitable
22　 for near-surface disposal.
23
24　 For a performance assessment and inadvertent intruder calculation supporting 10 CFR 61.58,
25　 the waste should be represented as it is physically expected to be present and not be averaged
26　 over the stabilizing and encapsulating materials unless the estimated doses to the public and
27　 inadvertent intruders would be conservative as a result of averaging. Otherwise, every attempt
28　 should be made to represent the expected distribution of activity within the disposal system. If
29　 the 10 CFR Part 61, Subpart C, performance objectives can be met with reasonable assurance,
30　 then the waste is considered to be acceptable for near-surface disposal.
31
32　 When performing the intruder calculations, it is not appropriate to calculate an average dose
33　 factoring in the likelihood of the occurrence of the scenario. The likelihood of the intruder
34　 scenario occurring is already represented in the higher limit {e.g., 5 mSv/yr [500 mrem/yr]}
35　 applied for inadvertent intruder regulatory analysis.
36
37　 Example—Other Provisions
38
39　 _A waste heel remains in a waste tank. The waste is well mixed with reducing grout using mixer_
40　 _pumps. A high-strength grout is placed over the reducing grout to provide a robust intruder_
41　 _barrier and to limit water contact with the waste. The top of the waste residuals is 10 m [33 ft]_
42　 _below the ground surface. The waste has a concentration of Tc-99 of 50 Ci/m³._
43

1 *Using a Category 1 approach the average concentration of Tc-99 in the wasteform is calculated*
2 *to be 6 Ci/m³ after mixing of the waste and grout; therefore the waste would be classified as*
3 *being greater than Class C (6 Ci/m³ is above the 3 Ci/m³ limit from Table 1 of 10 CFR 61.55).*
4 *DOE states that although the material is greater than Class C, the material is suitable for*
5 *near-surface disposal because the performance objectives in 10 CFR Part 61, Subpart C, can*
6 *be satisfied. DOE provides a performance assessment and intruder analysis for NRC staff*
7 *to review.*
8
9 *To demonstrate that the performance objectives can be satisfied, an intruder scenario is*
10 *evaluated in which a well driller places a well through the disposal system. In this case, the*
11 *acute intruder is exposed to drill cuttings (waste). The average concentration of the waste used*
12 *in the performance assessment calculations is calculated by assuming mixing over the volume*
13 *of well cuttings exhumed because the cuttings are expected to be well mixed when spread on*
14 *the land surface. The limiting dose impact to the intruder (well driller) from Tc-99 is calculated in*
15 *the intruder analysis to be 0.057 mSv/yr [57 mrem/yr] to a chronic receptor (a resident who*
16 *places a garden in the area contaminated by the drill cuttings). The waste is determined to*
17 *impose an acceptable risk to the inadvertent intruder.*
18
19 *Because the rate of erosion at the site is relatively high, a second intruder scenario is evaluated*
20 *in which most of the cover is eroded over the analysis time period. Some cover is expected to*
21 *remain. The intruder is assumed to construct a home in the area over the tank. Because the*
22 *direct exposure pathway is the only major contributing pathway for this scenario (the waste is*
23 *deeper than the excavation depth), the actual waste distribution is used in the performance*
24 *assessment. Alternatively, the average concentration of waste over the stabilizing materials*
25 *could have been used in the performance assessment because there would be less shielding*
26 *for this calculation and the doses would likely be conservative. The cover remaining provides*
27 *adequate shielding to reduce the dose from direct exposure to Tc-99 to insignificant levels;*
28 *therefore, the waste is determined to impose an acceptable risk to the inadvertent intruder.*
29
30 *The dose to a public receptor, who is offsite when institutional controls are in place and at the*
31 *edge of a buffer zone near the closed tanks after institutional controls end, is evaluated with an*
32 *all-pathways performance assessment. The performance assessment represents expected*
33 *degradation of the system over time. The modeling of the source term represents the waste as*
34 *two zones: one zone of higher hydraulic conductivity and reducing conditions that persist for*
35 *500 years and one zone of lower hydraulic conductivity and reducing conditions that persist for*
36 *the entire analysis period (10,000 years). The first zone represents the wasteform located at*
37 *the periphery of the tank. Because of shrinkage effects or degradation of the grout itself over*
38 *time from various attack mechanisms, waste in the first zone may be exposed to increased*
39 *moisture flow and higher rates of oxidation. The second zone represents waste that was*
40 *immobilized in the center of the tank in the reducing grout. The concentrations of radionuclides*
41 *in both zones are relatively homogeneous because the waste and grout were well mixed with*
42 *the mixer pumps. (If the distribution of waste was heterogeneous, then the waste should have*
43 *been represented in the performance assessment by the expected distribution of contamination*
44 *within the zones or distributions that could be demonstrated to be conservative with respect to*
45 *release and exposure modeling. The potential pathways of water to the waste may depend on*
46 *the discrete features of the system [e.g., cooling coils, shrinkage effects, fractures].) The*
47 *limiting dose impact to the public receptor from Tc-99 is calculated in the performance*
48 *assessment to be 0.012 mSv/yr [12 mrem/yr] to a chronic receptor (a resident who has a house,*

uses groundwater for domestic purposes, and grows produce in a small garden). The waste is
determined to impose an acceptable risk to the general population.

Disposal of the waste satisfies the performance objectives found in of 10 CFR Part 61,
Subpart C . After evaluating the specific characteristics of the waste, disposal site, and method
of disposal, the Commission finds reasonable assurance of compliance with the performance
objectives; therefore, the material is considered to be suitable for near-surface disposal.

3.5.1.2   Consultation for Disposal Plans for Waste Exceeding Class C

Waste with concentration limits above the concentration limits for Class C LLW as defined in
10 CFR 61.55, although generally unacceptable for near-surface disposal, may be acceptable
for near-surface disposal with special processing or design. The form and disposal methods for
this waste must be different and generally more stringent than those specified for Class C waste
[see 10 CFR 61.55(a)(2)(iv)]. If DOE determines that the waste exceeds the Class C
concentration limits or if DOE is unable to determine whether the Class C limits are exceeded,
the NDAA requires that disposal must be pursuant to plans DOE developed in its consultation
with the Commission (see Section 2.1).

The reviewer should take a risk-informed, performance-based approach when evaluating any
disposal plans DOE proposed during consultations. A risk-informed, performance-based
approach provides DOE with the flexibility to define disposal methods for wastes that do not
meet the Class C concentration requirements. In conducting its review, the staff should
consider the following aspects of DOE's plans:

- How DOE's disposal plans, with respect to form and disposal methods, are different and,
  in general, more stringent than plans that would be proposed for disposal of Class C
  waste; and

- How DOE demonstrates compliance with the performance objectives of 10 CFR Part 61,
  Subpart C (see Section 2 of this guidance). The review methods in Sections 4–7 of this
  guidance document can be applied to evaluate performance assessments and facility
  designs Intended to achieve compliance.

3.5.2   Review Procedures

The reviewer should evaluate the waste classifications provided in the waste determination
using the information in 10 CFR 61.55. Specifically, the reviewer should confirm the
following information:

- The appropriate tables in 10 CFR 61.55 have been used to classify the waste.

- Uncertainties in concentrations that are used to determine waste classification have
  been considered appropriately.

- Classification has been made based on the final wasteform(s).

- The sum of fractions has been used correctly (10 CFR 61.55), if applicable.

- The waste concentration averaging guidance (see Section 3.5.1.1) has been applied appropriately. Specifically, if concentration averaging is applied, the reviewer should confirm the following information:

  — If a Category 1 approach is used, residual waste and other materials are reasonably well mixed, and the type of averaging (volume- or mass-based) is consistent with the units of concentration specified for each radionuclide in Tables 1 and 2 of 10 CFR 61.55.

  — If a Category 2 approach is used, credit is taken only for material needed to stabilize waste rather than to stabilize a disposal facility (e.g., in most cases, the ratio of the unstabilized to stabilized radionuclide concentrations would not be significantly greater than a factor of 10 for waste classification purposes).

  — If a Category 2 approach is used, radionuclide concentrations in waste to which concentration averaging is applied are likely to approach uniformity in the context of applicable intruder scenarios.

  — Different waste streams have been classified separately, if appropriate.

  — If a Category 3 approach is used, site-specific, risk-informed approaches have selected intruder scenarios that are reasonably conservative and consistent with those discussed in Section 3.5.1.1.

  — If a Category 3 approach is used, adequate technical basis has been provided for the performance of intruder barriers for waste classification calculations.

  — If a Category 3 approach is used, the depth to waste is defined based on current and future projected disposal system configurations, and intruder scenarios consistent with the depth to waste have been selected for waste classification calculations.

  — Classification calculations using a Category 3 approach are adequately described and justified. Staff may use the example averaging expressions found in Section 3.5.1.1 as a review tool. However, the example averaging expressions are not to be used as the basis for classification calculations for the reasons discussed in Section 3.5.1.1.

  — Decay and ingrowth have been included in the classification calculations.

- In addition, if the waste exceeds Class C concentration limits, the reviewer should ensure that the consultation process described in Section 3.5.1.2 is applied appropriately.

3.6   References

Evans, M.S. "A Review of the EMMA Manipulator System with Regard to Waste Retrieval from Hanford Underground Storage Tanks." PIT–MISC–0129. White Paper for the Tanks Focus Area, Office of Science and Technology, DOE. 1997.

Gilbreath, K.D. "Risk Benefit Evaluation of Residual Heel Removal in Tanks 19 and 18." CBU–PIT–2005–00169. Rev. 0. Westinghouse Savannah River Company, Closure Business Unit, Planning Integration and Technology Department. 2005.

Hatchell, B., et al. "Russian Pulsating Mixer Pump Deployment in the Gunite and Associated Tanks at ORNL." PIT–MISC–0132. American Nuclear Society Paper No. 001. April 2001.

ICRP. "Report of ICRP Committee II on Permissible Dose for Internal Radiation." International Commission on Radiological Protection. 1959.

———. "ICRP Publication 60: 1990 Recommendations of the International Commission on Radiological Protection Annals of the ICRP Volume 21/1-3." April 1991.

Leishear, R.A. "ADMP Mixing of Tank 18F: History, Modeling, Testing, and Results." WSRC–TR–2004–00036. Rev. 0. Westinghouse Savannah River Company. 2004.

Office of Management and Budget. "Guidelines and Discount Rates for Benefit-Cost Analysis of Federal Programs." Circular A–94. 1992.

U.S. Department of Energy (DOE). "Innovative Technology Summary Report: Light Duty Utility Arm." DOE/EM–0492, OST Reference No. 85, Tanks Focus Area, Office of Science and Technology, DOE. 1998.

———. "Radioactive Waste Management." DOE O 435.1. August 2001.

———. "Draft Section 3116 Determination Idaho Nuclear Technology and Engineering Center Tank Farm Facility." DOE/NE–ID–11226. DOE, Idaho Operations Office. September 2005a.

U.S. Nuclear Regulatory Commission (NRC). "Draft Environmental Impact Statement on 10 CFR Part 61, Licensing Requirements for Land Disposal of Radioactive Waste." NUREG–0782. September 1981.

———. "Update of Part 61 Impacts Analysis Methodology." NUREG/CR–4370, Vol.1. 1986.

———. "Reassessment of NRC's Dollar Per Person-Rem Conversion Factor Policy." NUREG–1530. 1995a.

———. "Branch Technical Position on Concentration Averaging and Encapsulation." January 1995b.

———. "Regulatory Analysis Technical Evaluation Handbook." NUREG/BR–0184. January 1997.

———. "Technical Report on a Performance Assessment Methodology for Low-Level Radioactive Waste Disposal Methods." SECY–00–0182. April 2000.

———. "Decommissioning Criteria for the West Valley Demonstration Project (M–32) at the West Valley Site; Final Policy Statement." *Federal Register,* 67 FR 5003. February 2002.

1 ⸻. "Environmental Review Guidance for Licensing Actions Associated With NMSS
2 Programs. Final Report." NUREG–1748. August 2003a.
3
4 ⸻. "Consolidated NMSS Decommissioning Guidance." NUREG–1757. Vols. 1–3.
5 September 2003b.
6
7 ⸻. "Technical Evaluation Report for the U.S. Department of Energy, Savannah River Site
8 Draft Section 3116 Waste Determination for Salt Waste Disposal." Letter from L. Camper to
9 C. Anderson, DOE. Washington, DC. December 2005.
10
11 ⸻. "Regulatory Analysis." Circular A-4. 2003.
12
13 Sams, T.L. "Stage II Retrieval Data Report for Single-Shell Tank 241–C–106." RPP–20577,
14 Rev. 0. CH2M Hill Hanford Group, Inc. May 2004.
15
16 Schlahta, S.N. and T.M. Brouns, PNNL-11906, "Tanks Focus Area FY98 Midyear Technical
17 Review." Tanks Focus Area Technical Advisory Group, Office of Science and Technology,
18 DOE. 1998.
19
20 Vesco, D.P., et al. "Lessons Learned and Final Report for the Houdini Vehicle Remote
21 Operations at the Oak Ridge National Laboratory." American Nuclear Society Paper No. 083.
22 April 2001.
23

4 PERFORMANCE ASSESSMENT

This section provides guidance to review the performance assessment the U.S. Department of Energy (DOE) uses to evaluate dose for the nominal case (i.e., cases other than intruder scenarios) to demonstrate compliance with the performance objectives of 10 CFR 61.41. The term performance assessment can refer to (1) the process of estimating future radiation doses to receptors or (2) a model or collection of models (e.g., process or submodels) used to estimate future radiation doses to receptors. The review will encompass evaluation of scenario selection, the performance of the engineered system, the release and migration of radionuclides through the engineered barrier system and the geosphere, and radiation dose to the receptor groups. The review should be performed in a risk-informed manner, so that the reviewer is focused on those areas that have the largest impact on the estimated doses. Additional guidance on developing performance assessment and dose assessment models is found in NUREG–1573 (NRC, 2000) and NUREG–1757 (NRC, 2003). Protection of inadvertent intruders (10 CFR 61.42) is discussed in Section 5, protection of individuals during operations (10 CFR 61.43) is discussed in Section 6, and site stability (10 CFR 61.44) is discussed in Section 7 of this guidance document. Typically, a performance assessment is developed to demonstrate whether the performance objectives have been met. A performance assessment is a quantitative evaluation of potential releases into the environment and the resultant radiological doses. Abstraction describes the simplification of information in a performance assessment. The degree of abstraction in a performance assessment model (e.g., a highly abstracted, simplified model or a direct integration of complex process models) typically represents a balance between practical aspects (e.g., maintaining computational efficiency, allowing for efficient modification of the model, and ensuring the model can be relatively easily understood and evaluated) and preserving details that may significantly affect the results. In general, the complexity of the performance assessment should be commensurate with the amount of support to justify the results of the assessment.

This guidance document provides a consistent set of areas of review and review procedures to ensure uniformity of reviews performed for different sites by different review teams. However, the guidance document also affords flexibility to the reviewer to perform a more detailed review of particular elements of the performance assessment if justified by the risk significance of the elements (see Sections 4.2 and 4.6).

The performance assessment documentation will commonly justify the data used, describe the models used, verify and support the models, and evaluate the impact of data and model uncertainty. To evaluate uncertainty, a variety of techniques may be used, including deterministic analysis with sensitivity analysis and probabilistic analysis with uncertainty and sensitivity analyses. The sensitivity analysis results may be used to conduct a risk-informed evaluation through the in-depth review of those parameters and processes most important to system performance with respect to meeting the performance objectives.

In general, different approaches to performance assessment calculations (e.g., deterministic, probabilistic) have their advantages and disadvantages. A deterministic approach can be very valuable when the analysis is clearly conservative, because it makes the demonstration of meeting the performance objectives more straightforward, and it can be significantly easier to interpret results and explain them to stakeholders. While deterministic analysis can be a suitable methodology for performance assessment, it can also present a challenge when used to represent a system that responds in a highly nonlinear fashion to changes in the independent

1 variables. In addition, when numerous inputs (e.g., data or models) are uncertain, evaluating
2 the impacts of the uncertainties on the decision can be a challenge with a deterministic analysis.
3 A typical one-off type of sensitivity analysis (e.g., where a single parameter is increased or
4 decreased) will only identify local sensitivity within the parameter space, such that it may not
5 clearly identify the risk implications of the uncertainty in the parameter. A probabilistic approach
6 can have distinct advantages when there are a number of uncertainties that may significantly
7 influence the results of a performance assessment or when the interdependence of parameters
8 or assumptions is not clear (e.g., for highly nonlinear problems). However, there are limitations
9 to probabilistic analysis, such as limited data to define parameter distributions and inappropriate
10 impacts on the performance metric (e.g., peak mean dose) resulting from selection of overly
11 broad parameter distributions, particularly for parameters that affect the timing of doses. Even
12 with a probabilistic approach, conceptual model uncertainty may not be explicitly represented
13 and therefore could not be assessed with uncertainty analysis. This guidance document
14 provides the details to enable the review of deterministic or probabilistic analysis. Review
15 procedures specific to each type of analysis are provided, as well as guidance to help the staff
16 determine when a deterministic analysis may not be sufficient.
17
18 The term "conservatism," as used with respect to performance assessment, is a relative term.
19 Conservatism is typically defined with respect to what is known or sometimes with standard
20 practices that have yielded acceptable performance (e.g., a safety factor used in the design of a
21 bridge). The use of the term "conservatism" with respect to performance assessment is typically
22 more conjectural in nature. For example, a parameter value may not be measured, and
23 therefore the analyst will attempt to select a conservative value based on professional judgment.
24 If a large amount of data is available to support a performance assessment model, less
25 conservatism would be needed in the analysis. In this regard, model support (i.e., information
26 that supports the results of a model) of process model results plays a key role in developing
27 confidence in the output of performance assessment calculations. Because of the long time
28 periods for which system performance is being estimated, performance assessment models
29 cannot be validated in the traditional sense. However, multiple methods for developing
30 confidence in the model projections can be used, including laboratory experiments, alternative
31 modeling approaches, field measurements, natural analogs, and expert elicitation, among
32 others. The amount of model support provided should be commensurate with the risk reduction
33 the natural and engineered system provides. Multiple lines of evidence are particularly
34 important when the risk reduction of the systems being evaluated is large.
35
36 This section of the guidance document discusses scenario selection and receptor groups
37 (Section 4.1), general technical review procedures (Section 4.2), specific technical review
38 procedures for such areas as infiltration, engineered barriers, and source term (Section 4.3),
39 computational models and computer codes (Section 4.4), uncertainty/sensitivity analysis
40 (Section 4.5), evaluating model results (Section 4.6), and evaluating whether releases are as
41 low as reasonably achievable (Section 4.7).
42
43 During the technical review performed using the guidance in this section, the reviewer should
44 identify those factors that are important to assessing compliance with 10 CFR 61.41. By
45 applying this guidance, the reviewer should determine whether adequate support has been
46 provided for important assumptions, data, and models. Important factors may need to be
47 confirmed or verified during the monitoring process for a variety of reasons, such as use of
48 uncertain information or planned installation of an engineered system after the review. As
49 outlined in Section 10 of this guidance document, NRC will emphasize those factors during

1 monitoring. This guidance document ensures that the performance assessment is a robust
2 assessment of system performance. However, many uncertainties are associated with most
3 performance assessments, and monitoring is a mechanism to more effectively manage those
4 uncertainties. Monitoring can also evaluate new information that may confirm or refute previous
5 information. Monitoring can verify assumptions; however, the performance assessment must be
6 adequately supported to demonstrate compliance with the performance objectives. The main
7 elements within this section (system description, data sufficiency, data uncertainty, model
8 uncertainty, and model support) were specifically developed to ensure the performance
9 assessment is technically adequate. It is reasonable to expect that key assumptions have
10 adequate technical basis and that the assumptions can be verified during monitoring.
11 Monitoring is not to be used as a substitute for inadequate information, but rather a confirmation
12 of the previous determination of adequacy considering uncertainty. The reviewer should
13 consider the sensitivity and uncertainty analysis and barrier analysis information presented by
14 DOE and discussed in Sections 4.5 and 4.6, respectively. Information supporting the
15 performance of the natural and engineered systems that provide the largest risk reduction
16 should be noted and emphasized in monitoring, as appropriate.
17

18 ## 4.1 Scenario Selection and Receptor Groups
19

20 Scenario analysis is typically the initial step in the model development (or selection) process.
21 Evaluation and selection of applicable features and processes at the disposal site and the
22 surrounding region supports the conceptualization of the total system (i.e., disposal site and
23 surrounding area) and provides confidence in the completeness of the performance assessment
24 model. A scenario description should serve as a broad conceptual roadmap for the
25 performance assessment model, emphasizing those key features and processes that influence
26 the release, transport, and dose from radionuclides migrating from the disposal site. The
27 scenario description should provide sufficient information to understand the general spatial
28 domain and conceptualization of the performance assessment model including release
29 location(s), applicable radionuclide transport pathways, and the location and general
30 characteristics of the receptor(s).
31

32 ### 4.1.1 Areas of Review
33

34 This section focuses on ensuring that appropriate scenarios for radionuclide release,
35 transport, and exposure of a receptor group have been considered for evaluation in the
36 performance assessment. The process for developing scenarios to evaluate in the
37 performance assessment typically includes the following steps: (1) scenario identification,
38 (2) identification of relevant features and processes, and (3) development of receptor
39 characteristics. The period of performance evaluated will influence the development of
40 scenarios for the performance assessment.
41

42 #### 4.1.1.1 Period of Performance and Institutional Controls
43

44 Generally, a period of 10,000 years after closure is sufficient to capture the peak dose from the
45 more mobile, long-lived radionuclides and to demonstrate the influence of the natural and
46 engineered systems in achieving the performance objectives (NRC, 2000). However,
47 assessments beyond 10,000 years may be necessary to ensure (1) that the disposal of certain
48 types of waste does not result in markedly high impacts to future generations or (2) evaluate
49 waste disposal at arid sites with extremely long groundwater travel times. Periods of

1   performance shorter than 10,000 years are generally not appropriate for disposal facilities for
2   incidental waste, because of the larger fraction of long-lived radionuclides compared to a typical
3   commercial low-level waste (LLW) disposal facility. Presenting and understanding long-term
4   risk (e.g., greater than 10,000 years) can be an important part of performance assessment
5   analyses, even if those risks are not used to demonstrate compliance with the performance
6   objectives of 10 CFR Part 61, Subpart C.
7
8   The regulations in 10 CFR 61.59(b) specify that institutional controls may not be relied upon for
9   more than 100 years after closure. When 10 CFR Part 61was developed, it was envisioned that
10 LLW in a disposal facility would decay, in a maximum of 500 years, to activity levels that would
11 not pose a significant risk to an inadvertent intruder (given assumptions about waste distribution
12 and waste management practices). Significant quantities of long-lived isotopes would be limited
13 such that long-term risks to the public from disposal facility would be acceptable. In developing
14 10 CFR Part 61, NRC considered longer periods of institutional control in the Draft
15 Environmental Impact Statement (NRC, 1981). Assumptions about the persistence of
16 institutional controls in the international community were considered, and a series of public
17 meetings was conducted to get stakeholder input. The consensus among the stakeholders was
18 that it is not appropriate to assume institutional controls would last for more than a few hundred
19 years. The regulatory philosophy is that the engineered and natural system should afford
20 protection to the public. Because of the relatively large uncertainty associated with predicting
21 societal systems, total reliance on institutional controls of a site for extended periods of time
22 was not practical. For the commercial disposal of LLW, a number of requirements are imposed
23 to assure that institutional controls will be maintained for 100 years. The institutional controls
24 allow monitoring and maintenance of the disposal facility to be completed and also restrict
25 access to a disposal facility after closure.
26
27 **4.1.1.2   Scenario Identification**
28
29 It is generally acceptable to conduct performance assessment modeling based on scenarios
30 such as those described in NUREG–1757 (NRC, 2003, Vol. 2, Appendix I). Scenarios used in
31 the performance assessment should generally account for site-specific data and information
32 about the characteristics of the site and surrounding region (including local practices), potential
33 disruptive processes, and temporal behavior of the engineered and natural barriers. As
34 necessary, the scenarios that are evaluated in the performance assessment should be
35 constrained in a manner consistent with the relevant guidance provided in this document
36 (e.g., scenarios should be based on past, current, and projected future activities at the site).
37
38 Release and transport scenarios are likely to involve mobilization of waste by infiltrating water,
39 contamination of local groundwater and/or surface water, and subsequent use of contaminated
40 groundwater or surface water for domestic, agricultural, and recreational purposes by receptors.
41 Gaseous releases to air may also need to be addressed. Releases may occur at the source
42 location and perhaps at the receptor location (e.g., due to emanation from groundwater) at
43 some sites. Receptor characteristics and exposure scenarios may vary from site to site;
44 however, drinking water and agricultural food production (crops, livestock) commonly contribute
45 to radionuclide intake by many types of receptors. External exposure to contaminated soils and
46 inhalation of resuspended contaminants also are common exposure pathways. Recreational
47 use of surface water (e.g., fishing and swimming, including exposure to contaminated
48 sediments) may be an exposure pathway at some sites.
49

1   The reviewer should ensure that appropriate exposure pathways are included in the
2   performance analyses or that technical justification is provided to explain why certain pathways
3   may not be applicable for a particular site. For example, common exposure pathways include
4   ingestion, inhalation, and external exposures. Transport pathways may be excluded from
5   performance analysis if it can be demonstrated that there is either limited potential for
6   radionuclides to be released into a particular pathway, or the pathway is not viable (e.g., water
7   is not potable). Specific review guidance for factors that should be considered for the
8   inadvertent intrusion exposure scenario is discussed in greater detail in Section 5. Additional
9   guidance for reviewing protection of individuals during operations is provided in Section 6.
10
11   **4.1.1.3   Identification of Relevant Features and Processes**
12
13   Prior to conducting detailed technical reviews of the performance assessment, the scenario(s)
14   must be sufficiently described and documented to confirm that key features and processes have
15   been included in the overall system model. An acceptable dose assessment analysis need not
16   incorporate all the physical, chemical, and biological processes at the site. The reviewer should
17   ensure that the scope of the analysis and the level of sophistication of the conceptual models
18   are suitable for demonstrating compliance with performance objectives in 10 CFR Part 61,
19   Subpart C. Examples of features and processes that are commonly considered include air, soil,
20   groundwater, surface water, plant uptake, and exhumation by burrowing animals. Transport
21   and exposure pathways may be excluded from the performance assessment if it can be
22   demonstrated that there is either no potential for radionuclides to be released into a particular
23   pathway, or the pathway is not a viable transport or exposure pathway for the particular
24   scenario (e.g., groundwater is not potable). The following list of general features and processes
25   provides major elements that are expected to be considered when scenarios are developed for
26   evaluation in the performance assessment. These scenarios will be a function of the projected
27   human activities at the site, features and processes of the engineered system, and features and
28   processes of the natural system. Specific features and processes that should be considered
29   when developing submodels of the performance assessment are reflected in the review
30   procedures of the pertinent sections. Reviewers should consider whether the following features
31   and processes are relevant:
32
33   •   Human activities at the site, with emphasis on local practices, that could bring people in
34   contact with waste (e.g., water use, hunting, fishing, recreational activities such a
35   swimming and boating, habitation in dwellings, other unique activities that involve water
36   use or ground disturbance);
37
38   •   Features of the site that affect potential exposure pathways (e.g., groundwater quality
39   and aquifer productivity);
40
41   •   Frequency and magnitude of disruptive processes (e.g., seismic events, floods,
42   hurricanes) and their impact on the release of waste to the environment;
43
44   •   Location of surface water bodies such as streams and rivers in relation to the waste
45   disposal facility;
46
47   •   Features of the site meteorology affecting transport of airborne contaminants, including
48   stability class, wind speed, wind direction, temperature, and rainfall;
49

1 • Features of the disposal site that may influence the degradation of engineered systems
2 and the release of radionuclides from those systems [e.g., the process of fluctuation in
3 the shallow water table at the Savannah River Site (SRS) may influence oxidation of and
4 radionuclide release from the wasteform];
5
6 • Features and properties of the waste inventory, wasteform, and the facility design that
7 define the release rate of radionuclides from the disposal facility (e.g., the low hydraulic
8 conductivity of an intact cementitious wasteform may limit release of radionuclides to
9 diffusional processes);
10
11 • Features of the disposal facility that may influence the release of radionuclides from the
12 system (e.g., discrete pathways resulting from features of the system [such as sumps,
13 piping, or shrinkage of the wasteform within the disposal container]);
14
15 • Processes that influence the partitioning and mobility of the waste inventory (e.g., the
16 complexation of radionuclides by chelating agents in the waste);
17
18 • Processes that influence the ability of the wasteform to retain radionuclides
19 (e.g., seismically induced fracturing of cementitious wasteforms);
20
21 • Features of local flora that may affect the release of waste (e.g., deep rooting species
22 may reduce the effectiveness of an infiltration cap over time) or the uptake of
23 contaminants by humans and animals;
24
25 • Features of local fauna that may affect the release of waste (e.g., burrowing animals
26 may exhume waste from disposal areas) or the uptake of contaminants by humans and
27 animals (e.g., deer may access contaminated vegetation or water sources that are not
28 viable for human receptors);
29
30 • Physical and chemical properties of surface soils such as hydraulic conductivity,
31 porosity, moisture characteristic curve parameters, erodibility, and distribution
32 coefficients that may influence the process of water infiltration and the retention of
33 radionuclides (e.g., sorption processes, erosion rates) at the disposal site;
34
35 • Physical and chemical properties of the saturated and unsaturated zones that influence
36 water infiltration and sorption processes (e.g., soil type, mineralogy); and
37
38 • Features of the unsaturated zone (e.g., anthropogenic features such as abandoned
39 wells or natural features such as fractures) that result in discrete transport pathways or,
40 conversely, that will justify assuming porous flow.
41
42 The scenarios evaluated in the performance assessment should integrate the relevant major
43 features and processes at the site. The goal is to develop a set of reasonably anticipated
44 natural conditions, processes, and events and their impact on the engineered disposal system
45 to be represented in the site conceptual model. Processes affecting the long-term stability of
46 the disposal site should also be considered in the development of scenarios for the performance
47 assessment. For example, 10 CFR 61.13 provides specific processes that should be
48 considered, including erosion, mass wasting, slope failure, settlement of wastes and backfill,
49 infiltration through covers over disposal areas and adjacent soils, and surface drainage of the

1 disposal site. Guidance for reviewing analyses of site stability is provided in Section 7 of this
2 document. Disruptive processes such as erosion, seismic disruption, or other natural hazards
3 such as hurricanes, tornados, fires, or floods may affect the integrity of the disposal site and
4 constitute potential release mechanisms. The reviewer should appropriately consider such
5 processes and develop performance assessment scenarios for those hazards that cannot be
6 ruled out.

7

8 Specific review guidance for factors that should be considered for the inadvertent intrusion
9 exposure scenario is discussed in greater detail in Section 5. Additional guidance for reviewing
10 protection of individuals during operations is provided in Section 6.

11

12 **4.1.1.4   Receptor Characteristics**

13

14 Receptor characteristics may differ for air, groundwater, and surface water pathway scenarios
15 and the receptor group lifestyle habits (e.g., regional differences) may differ accordingly.
16 Receptor characteristics may also differ for onsite worker and public exposure scenarios and for
17 inadvertent intruder scenarios. The assumptions regarding receptor location and lifestyle habits
18 should be appropriately integrated into the performance model.

19

20 To demonstrate compliance with 10 CFR 61.41 or 10 CFR 61.42, a receptor should be
21 assumed to engage in residential, agricultural, or other activities that are consistent with
22 regional practices. The public receptor, protected by 10 CFR 61.41, is assumed to be located
23 at the boundary of the DOE-controlled area until the end of the active institutional control period,
24 at which point the receptor is assumed to move to the point of maximum exposure outside of the
25 disposal site. A receptor engaging in activities on the disposal site, rather than outside the
26 disposal site, is regarded as an inadvertent intruder protected by 10 CFR 61.42.
27 A disposal site consists of disposal units and a buffer zone [10 CFR 61.7(a)(2)]. For
28 convenience, the minimal area enclosing all of the disposal units in a disposal site (but not the
29 buffer zone) often is referred to as the "disposal area" (Figure 4-1). A buffer zone is a portion of
30 the disposal site that lies under the site and between the boundary of the disposal site and any
31 disposal unit. The buffer zone provides space to establish monitoring locations and to take
32 mitigative measures if needed [10 CFR 61.7(a)(2)]. An appropriate buffer zone is expected to
33 extend approximately 100 m [330 ft] from the disposal area in most cases. Alternate buffer
34 zone sizes may be acceptable if supported by technical justification. In general, appropriate
35 technical justification is expected to demonstrate that an alternate amount of space is needed to
36 implement monitoring activities or potential mitigative measures. Buffer zones should not be
37 enlarged solely to facilitate compliance with the performance objectives.

38

39 In the case of a tank farm, the tanks are expected to be regarded as disposal units and the
40 typical buffer zone is expected to extend approximately 100 m [330 ft] beyond any tank. In
41 some instances, the point of maximum exposure may be more than 100 m [330 ft] from the
42 disposal area. For example, a complex hydrogeologic system or overlap of plumes from
43 multiple sources may cause the maximum concentration of radionuclides in groundwater to
44 occur more than 100 m [330 ft] from the disposal area. In this case the location of the public
45 receptor protected by 10 CFR 61.41 would not be expected to coincide with the boundary of the
46 buffer zone.

47

48

1
2

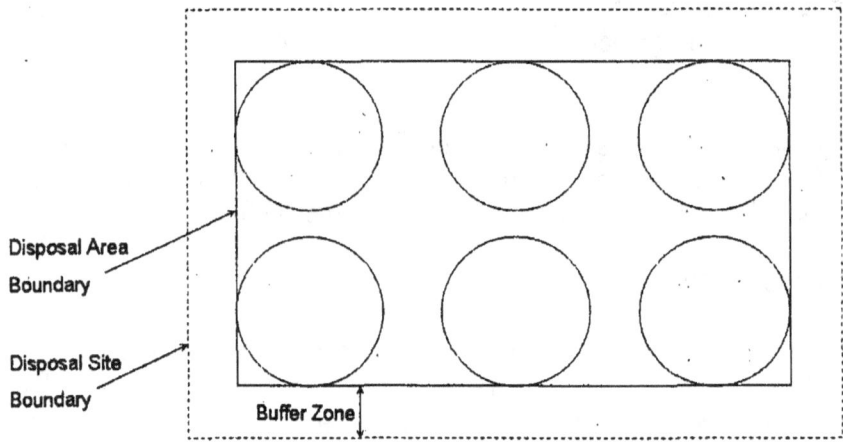

Each circle represents a disposal unit

Figure 4-1. Geometric Relationship of the Disposal Units, Disposal Area, and Buffer
Zone of a Disposal Site. Adapted From the Final Environmental Impact Statement for
10 CFR Part 61 (NRC, 1982).

4.1.2   Review Procedures

Details of the analysis related to the exposure pathways in the biosphere and dosimetry are
largely determined by the scenario and the assumed behavior of the receptor. Accordingly,
models related to the exposure pathways in the biosphere and dosimetry should not change
from one site to another unless there is a significant change in the scenario and associated
receptor behavior or location. In general, there are two primary areas of the dose analysis
where the conceptual model is expected to change from one site to another; these are related
to the source term (including the effects of engineered barriers on release) and environmental
transport. The reviewer should ensure that site-appropriate source terms and transport pathway
models are summarized in the conceptual model description. Source term analysis is
discussed in Section 4.3.3, and the principal environmental transport pathways (groundwater
[including transport through the unsaturated zone], surface water, and air) are discussed in
Section 4.3.4. The reviewer should perform the following procedures:

- Ensure that the scenarios used in the performance assessment to demonstrate
  protection of the general population from releases of radioactivity include radionuclide
  transport pathways via groundwater, surface water, soil (via erosion), and air, or that a
  sufficient technical basis is provided for their exclusion. Transport pathways may be
  excluded from performance analysis if it can be demonstrated that there is either limited

potential for radionuclides to be released into a particular pathway, or the pathway is not viable (e.g., water is not potable).

- Ensure that disruptive processes such as erosion or seismic disturbance and natural events such as hurricanes, tornados, fires, or floods have been appropriately considered in the scenario selection process. For example, erosion at a waste disposal facility may result in the need for development of performance assessment scenarios of waste transport via suspension in surface water or atmospheric pathways.

- Ensure that adequate technical basis has been provided for screening disruptive processes from representation in the performance assessment model and that disruptive processes which were not screened were implemented in the performance assessment models.

- Evaluate whether assumptions regarding receptor location and information defining human activities, such as land use and food consumption, are appropriately chosen for each exposure scenario and each exposure pathway.

- Determine whether lifestyle habits based on regional practices may be less conservative than common generic lifestyle habits. Adequate technical basis should be provided for locally defined lifestyle habits that are less conservative than commonly accepted lifestyle habits in generic dose assessments.

- Evaluate whether information for each scenario is presented in a clear and transparent manner (e.g., each pathway is listed, receptor characteristics and lifestyle behavior are described, maps showing potential receptor locations with respect to the disposal site are provided).

- Ensure that the public receptor location for evaluating compliance with 10 CFR 61.41 is at the point of maximum exposure outside the buffer zone (see Section 4.1.1.4).

- Ensure that the buffer zone has been appropriately defined, generally not to exceed 100 m [330 ft] from the disposal area. The reviewer should ensure that adequate technical basis is provided for a proposed buffer zone size significantly larger than 100 m. In general, buffer zone size is expected to be based on the amount of space needed to perform monitoring activities and any necessary mitigative measures.

- Ensure that DOE has considered applicable features and processes that could significantly influence disposal site performance in developing its scenarios (see Section 4.1.1.3).

- Ensure that DOE has not taken credit for active institutional controls for more than 100 years after closure.

- Ensure that the analysis estimated performance for a period of 10,000 years to demonstrate compliance with 10 CFR Part 61, Subpart C. A period of 10,000 years is generally sufficiently long to (1) evaluate performance of both engineered barriers and the site and (2) capture the peak doses from moderately mobile long-lived radionuclides.

- Determine, if appropriate, that DOE has evaluated doses beyond 10,000 years (e.g., for strongly sorbing radionuclides or very long-lived engineered barriers) and whether those doses are expected to be markedly higher than those evaluated for compliance with 10 CFR 61.41 (i.e., 10,000 years). If so, the reviewer should note in the Technical Evaluation Report the maximum doses expected to occur after 10,000 years.

- Determine whether DOE has considered waste inventory characteristics such as activity, half-life, and mobility of radionuclides in determining whether the appropriate performance period should be longer than 10,000 years.

- Ensure that processes and conditions that control engineered barrier degradation, water infiltration, leaching of waste, and release and transport of radionuclides to the general environment have been considered in development of scenarios for the performance assessment.

- Ensure scenario descriptions and any associated figures or maps depicting scenarios are supported by and consistent with the system descriptions and information regarding regional features and practices.

- Verify that scenario descriptions are consistent with the models implemented in the performance assessment.

- Ensure that DOE has considered the list of general features and processes provided in Section 4.1.1.3 in development of scenarios for implementation in the performance assessment.

4.2 General Technical Review Procedures

The review should focus on understanding the importance of various assumptions, models, and data in the performance assessment. As discussed in Section 4, the performance assessment model can be a collection of other models (e.g., submodels or process models) of varying levels of complexity, or it can be an integrated model. Regardless of the form of the numerical representation, the performance assessment model will represent numerous processes (e.g., infiltration, degradation of engineered barriers, release and transport of radionuclides, exposure of receptors to radionuclides). There are general technical review procedures that are applicable for all parts of the DOE performance assessment model. This section identifies those review procedures that can be broadly applied. Specific review procedures are identified in Section 4.3. The general technical review procedures can be divided into five separate categories:

- System Description: These review procedures ensure that DOE has adequately described the performance assessment models and the overall disposal system and that the different performance assessment models have been appropriately integrated (e.g., infiltration with source term release). The description should be adequate for the reviewer to understand the modeling and analyses, and if necessary, to perform independent analysis of the disposal system (see Section 4.2.1.).

- Data Sufficiency: These review procedures ensure that DOE has provided sufficient data to support the performance assessment models. The types of data to be considered may include site-specific data (e.g., laboratory measurements and field-scale measurements or experiments), data from analogous sites, data from generic sources, output from detailed process-level models, and expert judgment (see Section 4.2.2.).

- Data Uncertainty: These review procedures ensure that DOE has captured the variability in data and provided an assessment of uncertainty due to the incomplete knowledge of the natural system, engineered system, or inventory. Parameter uncertainty can be propagated through the performance assessment by distributions of variables (probabilistic analysis) (e.g., hydraulic conductivity, porosity, retardation coefficient). In a deterministic analysis, the data uncertainty can be examined with sensitivity analyses and bounded by the selection of conservative values (see Section 4.2.3.).

- Model Uncertainty: These review procedures ensure that DOE has evaluated the impact of model uncertainty and discussed the inherent uncertainties in applying predictive models: (1) over long periods of time for which direct validation is precluded and (2) to complex systems for which measurement and characterization may be limited. These uncertainties can be evaluated in the performance assessment by considering reasonable ranges in conditions and processes to test the robustness of the model, by using distributions of parameters to represent the likely ranges in conditions or processes, or by bounding the effects of model uncertainty by using conservative assumptions. Ideally, model uncertainty is minimized by developing as much model support as practical (see Section 4.2.4.).

- Model Support: These review procedures ensure that the output from the DOE performance assessment model can be supported by comparison to independent data. In general, using these review procedures, the reviewer should expect to evaluate multiple lines of evidence supporting the selected model (e.g., field tests or laboratory tests that provide a technical basis for selecting a certain release mechanism). In addition, the reviewer may conduct independent analyses for comparison of process model results, or the model results may be compared to analogous systems (see Section 4.2.5.).

To review the overall performance assessment, the reviewer should recognize that models DOE used may range from highly complex process-level models to simplified models, such as response surfaces or lookup tables. The reviewer should evaluate the adequacy of the model and the supporting technical basis, regardless of the level of complexity. The reviewer should determine whether uncertainties in the models and parameters are appropriately accounted for in the DOE performance assessment. Specifically, the reviewer should follow the procedures given in Sections 4.2.1–4.2.5.

4.2.1   System Description Review Procedures

- Examine the descriptions of design features (including engineered barriers, wasteforms, and other engineered components) and the relevant natural system features (including the geological, hydrological, and geochemical aspects of the natural barriers at the site). Verify that the descriptions sufficiently support the development of a conceptual model of the site, including major pathways for water and radionuclide movement.

- Assess whether the design and natural system features have been adequately incorporated into the performance assessment. Where simplifications are used, confirm that the technical bases used to support the simplifications (e.g., modeling assumptions and approximations) are adequate (e.g., verify that the potential effects of the simplifications on dose predictions have been bounded) and have been documented in a transparent and traceable manner.

- Determine whether the conditions and assumptions used in the performance assessment modeling are consistent with the design feature documentation.

- Verify that the assumptions, data, and models DOE used are consistent among the different parts of the performance assessment. For example, the release models used in the source-term model should be consistent with the chemical environment assumed for the engineered barrier system.

- Confirm that common boundary and initial conditions are consistent among submodels of the performance assessment (e.g., the recharge rate in saturated zone modeling should be consistent with infiltration applied to the unsaturated zone if there is not significant lateral flow in the unsaturated zone).

- Examine how features and processes related to engineered and natural barrier system performance have been included in the performance assessment model, and verify that performance through time has been adequately represented. For example, if the hydraulic conductivity of a barrier is expected to increase over time, verify that the hydraulic conductivity value(s) used in the performance is consistent with or bounds the expected degradation.

- Evaluate whether conceptual models sufficiently account for the most important physical, chemical, and biological processes at the site so that a more realistic representation of the site would not lead to higher dose estimates.

4.2.2   Data Sufficiency Review Procedures

- Confirm that the data used to support conceptual models, process-level models, and simplifications in the performance assessment are sufficient. Examine the parameters used for these models, and verify that the parameters are based on adequate technical basis, such as data derived from laboratory experiments, site-specific field measurements, operational experience, research at comparable sites, and process-level modeling.

- Confirm that DOE has provided sufficient data on the characteristics of the waste, engineered barriers, and natural system to establish initial and boundary conditions for models.

- Verify whether sufficient data have been collected to adequately support modeled degradation of engineered barriers and near-field transport of radionuclides as well as to establish important characteristics of the natural system (e.g., geochemistry, hydrology).

- Verify that parameter values are derived from site-specific data when available or that an analysis is included to show that data from generic sources leads to a conservative assessment of performance.

- Verify that data from generic sources are appropriate for the site-specific conditions or materials in the performance assessment. For example, if distribution coefficients for cement are taken from Bradbury and Sarott, verify that the formulations used in the reference are consistent with the grout being used at the site (Bradbury and Sarott, 1995).

- Verify that experimental conditions for laboratory or field measurements (e.g., temperature, chemistry of a solution used in a leach test) are reasonably representative of expected system conditions or that an adequate assessment of the impact of the differences in conditions has been provided.

- Confirm that parameter values used in process-level models are appropriate for the time- and space-scales of the performance assessment calculations.

- Examine the initial and boundary conditions of the models, and verify that they are consistent with available data.

- Confirm that sensitivity or uncertainty analyses have been used to assess data sufficiency. If the analyses identified significant impacts associated with the uncertainty in particular data, evaluate additional data that was acquired to limit the uncertainty in results to an acceptable range. As appropriate, document the associated assumption in the TER.

- If expert judgment is used as a basis for selecting parameter values where default model parameters or site-specific data are not sufficient, evaluate the methods DOE used to develop the information. Confirm the information was developed from unbiased sources in a transparent and objective way [for example, see guidance in NUREG–1563 (NRC, 1996)].

- As appropriate, verify that DOE used acceptable approaches for peer review and data qualification [see guidance such as NUREG–1297 and NUREG–1298 (NRC, 1988a,b)], or provided adequate justification for using alternative approaches.

4.2.3   Data Uncertainty Review Procedures

- If deterministic models are used in the performance assessment, evaluate the technical bases for parameter values, assumed ranges used in sensitivity analyses to characterize data uncertainty, and bounding values used in conceptual and process models.

- If a deterministic approach is used, verify that key parameter values are reasonably conservative, technical basis is provided for conservative assumptions, and the conservatism of values is defined on a total system level and not at the local level. For example, increasing the hydraulic conductivity of saturated zone aquifers to address uncertainty may be conservative with respect to contaminant travel time but may be nonconservative with respect to dose as a result of increased dilution of contaminant fluxes entering the saturated zone from the unsaturated zone.

- For deterministic or probabilistic approaches, evaluate whether statistical correlation between parameters has been appropriately considered in the analysis. For example, a scenario of increased infiltration may necessitate a positive correlation (with respect to infiltration) of parameters related to the timing of oxidizing conditions in a wasteform and the timing of degradation of a steel barrier if the model is abstracted and is not representing the underlying physical processes. Verify that adequate technical basis or a bounding argument is provided for unanalyzed correlations.

- If probabilistic models are used in the performance assessment, evaluate the technical bases for parameter ranges, probability distributions, or bounding values. The reviewer should verify that the technical bases adequately support the treatment of uncertainty and variability of these parameters.

- Verify that uncertainty in initial and boundary conditions has been appropriately considered and is reflected in the performance assessment models.

- Confirm that uncertainty in data from both temporal and spatial variations has been incorporated into the parameter ranges (e.g., degradation of barrier performance with time, spatial variation of soil properties).

- Determine whether expert judgment was used as a basis for data uncertainty, and confirm the information was developed in a transparent and objective way [e.g., see guidance in NUREG–1563 (NRC, 1996)].

-

4.2.4 Model Uncertainty Review Procedure

- Compare the models used in the performance assessment with available data such as design data and verification tests for engineered barriers and wasteforms, laboratory experiments, field measurements, and monitoring data. Confirm that models in the performance assessment were developed considering the uncertainty and variability in supporting information.

- Verify that conceptual model uncertainties are adequately described and documented. Verify that the impact of model uncertainty on overall system performance was properly assessed.

- Examine the mathematical models included in the performance assessment. Evaluate the assumptions, the limitations, and uncertainties of the models and the bases for excluding alternative models.

- Verify that the models used in the performance assessment adequately represent or bound the uncertainty associated with underlying process-level models, if applicable. Where appropriate, use a detailed auxiliary analysis (i.e., an analysis performed outside of the overall performance assessment analysis) to verify that the DOE performance assessment approach reflects or bounds the uncertainties in the process-level models.

- Verify that the selected conceptual model is conservative or more representative of available supporting information relative to alternative models. Alternative models considered should (1) be reasonably consistent with available information or (2) should have been observed at other sites and cannot be refuted with existing information.

- Verify that quantitative evaluation of model uncertainty included the impact of data uncertainty in alternative models.

4.2.5   Model Support Review Procedure

- Evaluate the output from the performance assessment, and verify that DOE has compared the results with an appropriate combination of site characterization and design data, process-level modeling, laboratory testing, field measurements, analogs, and formal independent peer review.

- Examine the output from the mathematical models for consistency of the mathematical model response with the response expected based on the description of the conceptual models.

- Verify that the performance assessment model is reasonably supported by observations from the site, if available. For example, compare the output from the DOE performance assessment with inferences about fate and transport of radionuclides in the environment developed from data for environmental monitoring of leaks and spills.

- Where appropriate, use independent analyses to evaluate selected parts of the DOE performance assessment model, and assess whether the resulting doses are comparable.

- If possible, perform simplified calculations of processes, and compare to the intermediate outputs of the performance assessment models. For example, estimate the groundwater travel time of select radionuclides using information on hydraulic gradient, hydraulic conductivity, soil porosity, soil density, and radionuclide distribution coefficients, and compare to the travel times generated with performance assessment models.

- Confirm that DOE has identified and implemented adequate procedures to construct and test its mathematical and numerical models.

4.3   Specific Technical Review Procedures

In contrast to the general review procedures presented in the previous section, this section details review procedures that are specific to the different technical areas comprising component models of the DOE performance assessment. These review procedures were developed from experience in prior reviews and may be enhanced, modified, or supplemented based on future experience. As previously discussed, a performance assessment model may be a manually integrated or fully automated collection of individual models representing specific technical areas and processes. The individual models are commonly referred to as component models, process models, submodels, or abstractions. This terminology is used interchangeably in the specific review procedures that follow. The submodels are presented in a "top-down" sequence that is similar to that described in NUREG–1573 (NRC, 2000). Not all of the specific technical review procedures will apply to every waste determination, and the level of review

1   should be adjusted to reflect the significance of a given component (e.g., infiltration, engineered
2   barriers) to system performance (see Section 4.6). The topics included in the section are not all
3   inclusive; technical areas not covered should be reviewed in accordance with the general
4   technical review procedures.
5
6   It is expected that DOE will have a fairly large collection of technical reports covering a variety of
7   activities that have occurred at the DOE sites over the years. For example, hydrogeologic
8   studies and models may have been completed or developed to manage existing contamination
9   or to evaluate the performance of other disposal facilities for radioactive waste at the DOE site.
10   To the extent practical, the reviewer should consider other sources of information that may
11   support or refute the models and analysis used in the waste determination. For example, the
12   observed transport of Cs-137 from a LLW disposal facility at the DOE site may support or
13   refute the use of various values of K_d from a generic literature source used in the
14   performance assessment.
15
16   **4.3.1   Climate and Infiltration**
17
18   **4.3.1.1   Areas of Review**
19
20   This section focuses on the models and data that support climate projections, infiltration, and
21   unsaturated zone flow estimates used in performance assessments supporting waste
22   determinations. Temporal and spatial variations in processes and parameters related to
23   climate, infiltration, and unsaturated zone flow are potentially important to system performance
24   because one of the primary mechanisms of radioactivity release from a disposal facility is
25   leaching with water. A distinction between infiltration (that part of precipitation which moves
26   past the root zone) and water flow in the unsaturated zone (deeper flow whether through soil,
27   rock, or anthropogenic fill material) must be made. Infiltration can affect engineered barrier
28   performance and the release of radionuclides (see Section 4.3.2). Unsaturated zone flow is a
29   function of the physical characteristics of the unsaturated zone and the rate and distribution of
30   infiltration input to the unsaturated zone (e.g., boundary and initial conditions). Physical
31   characteristics of the unsaturated zone may be affected by anthropogenic and natural features.
32   Unsaturated zone flow affects the transport of radionuclides to the saturated zone.
33
34   4.3.1.1.1   Current Meteorology and Precipitation at the Site
35
36   The amount of water that contacts the waste is important for estimating the release of
37   radionuclides from a waste disposal facility. Knowledge of local meteorology and precipitation
38   is necessary to estimate infiltration and the potential for water to contact the waste. Current
39   information provides a baseline against which to evaluate the significance of any potential
40   changes over the period of performance (see Section 1.1.3).
41
42   Meteorological information is typically an input to calculations or models used to estimate
43   infiltration. The reviewer should evaluate information on precipitation (duration, intensity,
44   frequency, and seasonal variations of precipitation events), local air temperatures (daily and
45   seasonal variations), wind speeds and directions, and air humidity levels.
46
47   4.3.1.1.2   Current Infiltration and Unsaturated Zone Flow at the Site
48
49   The disposal horizon at most sites is located above the local water table in the hydrologically
50   unsaturated zone, sometimes also referred to as the vadose zone. Precipitation at a site can

1 follow several paths. Depending on the topography of the site and the composition of the
2 topmost soil or cover materials, water may tend to pond in some areas or to run off. Some
3 water may penetrate only the topmost layer of soil where it may be evaporated directly to the
4 atmosphere or be taken up by plants and then transpired back into the atmosphere.
5
6 Some fraction of the precipitation may move below the zone of evapotranspiration and contact
7 engineered barriers or other portions of the disposal system. This fraction is infiltration. Flow of
8 infiltration through the unsaturated zone will be affected by heterogeneities, fractures, and
9 anthropogenic changes and features that may lead to faster and preferred water pathways
10 (e.g., abandoned boreholes and wells). The amount of water that moves to the disposal
11 horizon, the frequency of the flow, and the spatial distribution of the flow are all potentially
12 significant to waste disposal facility performance. If waste is located below the water table, then
13 saturated zone flow would control the amount of water contacting the waste. Water contact may
14 be limited by the engineered barriers and related engineered systems.
15
16 The reviewer should evaluate the information on local soils and rocks that affect infiltration and
17 unsaturated zone flow (e.g., hydraulic conductivity, porosity, moisture content), vegetation
18 (types, distributions, seasonal changes), topography, erosion, runoff and drainage, and the
19 potential for flooding or ponding. The reviewer should examine the seasonal variation in the
20 independent variables for modeling infiltration and resultant unsaturated zone flow. For
21 example, in colder climate sites, a significant fraction of annual infiltration may result from snow
22 melt or similar processes when evapotranspiration is low. For modeling unsaturated zone flow,
23 DOE should provide properties (e.g., moisture characteristic curve parameters) that are
24 supported by empirical measurement or are sufficiently conservative (if generic information is
25 used). The reviewer should examine information provided on the spatial variability of features
26 and properties or the approach to address spatial variability. The reviewer should evaluate the
27 model support for estimates of unsaturated zone flow. Because unsaturated zone flow is
28 generally inherently more uncertain than saturated zone flow, a commensurate increase in the
29 model support or the conservatism of the analysis should be expected. The reviewer should
30 consider that spatial variability in features or hydrologic properties can affect the importance of
31 temporal variability. For example, the presence of fractures could increase the importance of
32 considering individual storm events in modeling long-term infiltration.
33
34 Information should be provided regarding the potential for perched zones to affect flow and
35 transport. If perched zones affect flow and transport, monitoring data should support the
36 location, extent, and persistence of the perched zones. Changes in operation that may affect
37 perched zones (e.g., use of percolation ponds) should be evaluated. If the perched zones are
38 important, the reviewer should determine whether DOE has evaluated the relative contribution
39 from both natural (e.g., recharge) and anthropogenic sources to the extent of the
40 perched zones.
41
42 4.3.1.1.3   Projected Meteorology and Precipitation at the Site
43
44 The reviewer should evaluate the current information on the meteorology and precipitation at
45 the site. The reviewer should also evaluate paleoclimatic information for the site. Recent and
46 current climate data are the best available predictors of the near-term future conditions at the
47 site whereas paleoclimate data may provide a basis for interpreting potential future changes in
48 the climate. It is important to assess the DOE assumptions to extrapolate the past and current
49 information and project those values and patterns into the future. For example, climate changes
50 may be assumed to be cyclical or linked to orbital patterns (e.g., Milankovitch forcing). Climate

1 projections should cover the full duration of the performance period. The reviewer should
2 examine how uncertainties inherent in projections of future climates have been accounted for
3 and how those uncertainties have been propagated through the performance assessment,
4 as appropriate. Sensitivity of system performance to the affects from the timing of climate
5 change (e.g., infiltration rates, moisture contents, water table depth, fluvial erosion) should be
6 evaluated to identify whether climate change induced by human activities can significantly affect
7 the performance assessment results. In many cases for buried waste, the impact of climate
8 change is mitigated by the damping function of the overlying geologic materials. In addition, a
9 proper consideration of the variability in hydrologic conditions for the present day may
10 encompass many future climate states. However, because predictive capabilities for future
11 climates continue to evolve, if a significant sensitivity is identified, staff should incorporate
12 current information regarding climate change, as appropriate, into the review.
13
14 4.3.1.1.4   Projected Infiltration and Unsaturated Zone Flow at the Site
15
16 In general, site conditions are expected to change over long time periods and to result in
17 changes to infiltration at the site. Changes to infiltration will produce changes in the unsaturated
18 zone flow. The reviewer should evaluate the available information on present day infiltration at
19 the site that may be used to project future infiltration rates. Projections of future infiltration
20 should cover the full duration of the performance period. Infiltration projections should account
21 for any construction or engineered features (see Section 4.3.2) that are designed to control or
22 reduce infiltration (e.g., caps, drainage layers, geosynthetics) or other changes to site conditions
23 that may affect infiltration (e.g., variations in precipitation, vegetation, or soil cover caused by
24 erosion). The reviewer should evaluate the integration of infiltration and water flow in the
25 unsaturated zone, if computed by different models. In reviewing barriers used to control
26 infiltration, the reviewer should evaluate any credit taken for maintenance or long-term
27 performance of these barriers after the loss of institutional controls. The relationship between
28 projected precipitation and projected infiltration should be described. Infiltration projections
29 should take into account the uncertainties inherent in such estimates and should propagate
30 those uncertainties through the performance assessment, as appropriate. Infiltration at
31 semiarid sites is generally controlled by short-duration (i.e., hourly or daily) storm events. Thus,
32 estimates of infiltration based solely on long-term (i.e., monthly or yearly) precipitation and
33 evaporation rates at a semiarid site may be misleading.
34
35 **4.3.1.2   Review Procedures**
36
37 To review this performance assessment submodel, the reviewer should consider the degree to
38 which DOE relies on climate and infiltration to demonstrate compliance with 10 CFR 61.41,
39 considering available sensitivity analyses, uncertainty analysis, and barrier or component
40 analysis (see Section 4.6). For example, the reviewer should perform a detailed review of this
41 area if DOE relies on estimates of infiltration that are significantly lower than natural recharge
42 values and that correspondingly produce lower release rates and provide significant delay in the
43 transport of radionuclides. If, on the other hand, DOE demonstrates that this submodel has a
44 minor impact on release rates or the transport of radionuclides to the receptor, then a simplified
45 review focusing on whether the analysis has been appropriately implemented should be
46 conducted. In general, higher infiltration rates will result in higher dose estimates, because
47 there will be higher mass flux rates of contaminants to the saturated zone. The overall dilution
48 in the calculation will be dominated by the saturated zone modeling and not the dilution in the
49 unsaturated zone (which increases with increased infiltration). In a risk-informed,

1 performance-based review, some of the following review procedures may not be necessary
2 when conducting a simplified review for models with a minor impact on performance:
3
4 • Apply the general review procedures found in Section 4.2 to the assessment of climate,
5 infiltration, and unsaturated zone flow.
6
7 • Confirm that an adequate baseline for current meteorology and precipitation at the site
8 has been used. Evaluate the adequacy of information for precipitation (duration,
9 intensity, frequency, and seasonal variations of precipitation events), local air
10 temperatures (daily and seasonal variations), wind speeds and directions, and air
11 humidity levels.
12
13 • Evaluate engineered features (see Section 4.3 of this guidance) that are designed to
14 control or reduce infiltration (e.g., caps, drainage layers, geosynthetics) or other changes
15 to site conditions that may affect infiltration (e.g., variations in vegetation or erosion of
16 soil cover). Evaluate credit taken for maintenance during the institutional control
17 period and long-term, passive performance of these barriers after the loss of
18 institutional controls.
19
20 • Examine the temporal and spatial relationship between infiltration and the water table
21 elevation for those disposal sites with waste located near the water table.
22
23 • Evaluate information regarding the potential for perched water zones to affect flow and
24 transport. Activities that may affect perched zones (e.g., use of percolation ponds)
25 should be evaluated. If perched zones are important to the performance of the site,
26 determine whether DOE has adequately evaluated the relative contributions from both
27 natural and anthropogenic sources to the extent of the perched zones.
28
29 • Verify that the data for infiltration are at appropriate time- and space-scales. Confirm
30 that adequate site-specific climatic, surface, and subsurface information is used.
31
32 • Confirm that precipitation estimates are based on long-term precipitation data that
33 adequately represent conditions at the disposal facility location. Long-term data for
34 precipitation are typically considered to extend over a period of several decades to
35 100 years.
36
37 • If estimates of infiltration are based on modeling, verify that the analysis has considered
38 seasonal variation in independent variables and short duration, large magnitude events,
39 especially when discrete high-permeability pathways are present in the near surface
40 (e.g., dessication cracks in a clay soil). The importance of temporal variability may be
41 dependent on spatial variability.
42
43 • Where applicable, confirm that adequate representation of the effects of fracture
44 properties, fracture distributions, matrix properties, heterogeneities, time-varying
45 boundary conditions, evapotranspiration, depth of soil cover, and surface-water runoff
46 and run-on is incorporated in the model or calculation.
47
48 • Confirm whether uncertainty in data, because of both temporal and spatial variations in
49 conditions affecting climate and infiltration, is incorporated into the selection of
50 deterministic parameters or the definition of parameter ranges. For example, the

reviewer should evaluate the climatic and hydrostratigraphic parameters used in the model to verify that they are consistent with site characterization data. The reviewer should confirm the data is sufficiently detailed such that heterogeneities that may influence the distribution and rate of liquid-water flux that has moved beyond the zone of evapotranspiration (infiltration and unsaturated zone flow) have been captured.

- Evaluate the assumptions DOE used to extrapolate from past climate data to future climate conditions. For example, the reviewer should determine whether climate changes are assumed to be cyclical or are linked to orbital patterns (e.g., Milankovitch forcing).

- Verify whether climate projections cover the full duration of the performance period and determine whether uncertainties in the projections are adequately accounted for and propagated through the performance assessment.

- Confirm that the performance assessment incorporates the hydrologic effects of future climate change that could alter the rates and patterns of present-day infiltration into the unsaturated zone.

- Evaluate the sensitivity of the performance assessment to the timing of climate change impacts (e.g., infiltration, saturation, water table depth, fluvial erosion) as a means to evaluate anthropogenic processes affecting climate.

- Ensure that infiltration estimates are either chosen in a clearly conservative manner or are supported by multiple lines of evidence. Typically, higher infiltration rates are more conservative, although in some circumstances a higher infiltration rate could result in a lower dose (e.g., dilution in perched water zones).

- Ensure that appropriate model support is provided for infiltration rates. For example, ensure infiltration rates are consistent with calibrated recharge rates from large-scale or regional flow models, other calculated values, infiltration rates from other site estimates, and values estimated based on soil properties.

- Because some DOE sites can be quite large, if site-specific infiltration rates developed or measured from other onsite areas are used to support the estimates for the waste disposal facility, confirm that they would be expected to be reasonably representative of local estimates for the waste disposal facility based on similarity of important variables (e.g., soil type, topography, vegetation).

- Ensure that the estimates of infiltration and unsaturated zone flow have appropriately accounted for anthropogenic features or actions at the site. For example, an undisturbed soil profile may have an infiltration rate that is different from that for a disturbed soil. Abandoned wells or other manmade features may act as discrete pathways for infiltration.

- Ensure that the parameters DOE used to model unsaturated zone flow are supported by empirical measurements or that generic information sources are sufficiently conservative. Confirm that DOE considered spatial variations in the properties of materials when selecting parameter values.

1 • Determine whether the impact of fractures or other naturally occurring discrete pathways
2 (such as sand lenses) have been represented in modeling flow in the unsaturated zone,
3 if applicable. Ensure that the selected model has not underestimated the potential
4 effects of preferential pathways.
5
6 • Determine whether adequate model support is provided for unsaturated zone flow
7 (e.g., field-scale observations or measurements, evaluation of the transport of past leaks
8 or spills) consistent with its risk significance. If a unit gradient approach is not adopted,
9 ensure that adequate model support is provided to justify a less conservative approach.
10

11 ## 4.3.2 Engineered Barriers

12
13 A wide variety of engineered barriers may be employed for incidental waste disposal, depending
14 on the nature of the waste and the planned disposal environment. Engineered barriers are
15 anthropogenic structures or devices intended to improve the disposal facility's ability to meet the
16 performance objectives in 10 CFR Part 61, Subpart C (10 CFR 61.2). In this document, the
17 term "engineered barrier" includes those anthropogenic barriers such as tanks, vaults, and other
18 components and systems that limit release of waste to the accessible environment (e.g., grout,
19 infiltration caps, erosion protection covers, slurry walls) or limit inadvertent intrusion into the
20 waste. In the performance assessment modeling, DOE may decide not to take credit for all
21 engineered barriers present at the site. Each type of engineered barrier will have a timeframe
22 over which it will be designed to perform its intended functions (e.g., the design life), which
23 should be justified for the specific application of the barrier. There is significant uncertainty in
24 the ability of engineered barriers to achieve the design goals when modeled performance is
25 needed for long periods of time, and the uncertainties tend to increase with increasing
26 performance periods. An engineered barrier with design goals significantly exceeding relevant
27 experience (either in degree or duration) should have a commensurately higher amount of
28 model support that the barrier will likely achieve the design goals. Regardless of the model
29 support, analysis should be performed to understand the impacts if the barrier does not achieve
30 its design goals.
31

32 ### 4.3.2.1 Areas of Review

33
34 This section focuses on the engineered barriers DOE proposed in its waste determination and
35 performance assessment. Improvement in the disposal facility performance can be achieved by
36 limiting the amount of water that contacts the wasteform, reducing the transport of radionuclides
37 within and from the site, and providing shielding from direct exposure, among other functions.
38 In particular, the review should focus on those aspects of the engineered barriers that are most
39 critical to meeting the performance objectives in 10 CFR Part 61, Subpart C. Section 4.6
40 provides guidance to evaluate the risk significance of barriers in the context of estimated
41 disposal facility performance.
42

43 4.3.2.1.1 Features and Dimensions of the Engineered Barrier System(s)

44
45 The reviewer should evaluate the descriptions of the engineered barriers proposed for the site.
46 The description of the engineered barriers will typically include the geometry, dimensions,
47 materials, functionality, design goals, and pertinent degradation mechanisms. Engineered
48 barriers may be above grade or below grade and may be designed for physical (e.g., vaults,
49 covers, erosion control barriers, drainage systems, containers, backfill, or infill) or chemical
50 control (e.g., pH buffers, oxygen getters). Radionuclide mobility through engineered barriers

1   may be affected by both the physical state of the barrier (e.g., low permeability and porosity)
2   and chemical phenomena such as sorption, precipitation, coprecipitation, dissolution, and ion
3   exchange. The reviewer should evaluate the potential for the physiochemical conditions
4   produced by the barriers to limit radionuclide mobility (e.g., by limiting flow and maintaining
5   reducing water compositions).
6
7   The reviewer should examine the design of the waste disposal system. The design should
8   specify the dimensions, orientation, and compositions of the engineered barriers. Specifically,
9   the reviewer should examine figures and illustrations (e.g., cross sections that illustrate the
10 components of the engineered barrier system). Those portions of the design for which DOE
11 takes credit as engineered barriers should be identified. The reviewers should examine the
12 design functionality (e.g., limit water contact with the waste, limit erosion) and properties of the
13 engineered barriers (e.g., porosity, hydraulic conductivity, sorption coefficients). The reviewer
14 should evaluate the design goals and the description and analysis of pertinent degradation
15 mechanisms to verify that the engineered barrier will likely be able to achieve the design goals.
16

17 4.3.2.1.2   Performance of Engineered Barriers

18
19 The effectiveness of engineered barriers, such as engineered caps and reducing grouts, is
20 expected to diminish over long time periods. Combinations of physical and chemical processes
21 will result in changes to the original barriers that may reduce their effectiveness (e.g., formation
22 of cracks in grout, concrete, or clay). The reviewer should examine the assumptions of barrier
23 degradation and the justification and technical bases for the time period for which DOE takes
24 credit for the effectiveness of the barriers.
25
26 For engineered barriers such as engineered caps that are designed to reduce infiltration
27 through the wasteform, the reviewer should evaluate the technical basis used to support
28 estimates of physical durability with time. Conceptual models (e.g., of fracturing of a cap or
29 wasteform that may affect the physical durability) should be supported by test results that are
30 appropriate for the materials to be used in the barrier. Because of the long time periods
31 involved, the reviewer should also evaluate information provided on the impact of biointrusion
32 (e.g., root penetration, burrowing animals) on engineered barrier performance. To estimate
33 evapotranspiration cover performance, it is important to consider processes that can affect the
34 ability of the plant community to transpire water, including but not limited to variation in weather
35 conditions, biotic processes, the formation of discrete pathways, disease, land use, and
36 disruptive events such as fires. The storage capacity of the cover should be designed
37 considering the performance period and requirements for moisture management at the site. In
38 addition, the dynamics of vegetation succession on the engineered cover should be considered
39 for the period of performance.
40
41 The reviewer should evaluate the technical basis provided to support DOE estimates of the
42 chemical performance of engineered barriers and wasteforms (e.g., reducing grout, saltstone)
43 during the performance period. For example, for a reducing cementitious wasteform the
44 reviewer should examine the analysis of the effects of temporal changes in pH and redox in
45 engineered materials (such as concrete) on source term K_d values, which may determine the
46 release rates of radionuclides. The persistence of the chemical durability of a barrier may be
47 directly related to the physical properties of the barrier. Information should be provided for the
48 reviewer to evaluate the coupling of physical and chemical degradation mechanisms of barriers.
49

1 For erosion control barriers (see Section 7), the reviewer should consider rock durability,
2 gradation, cover design, stability calculations for the top slope, side slope, and apron for any
3 cover, and other considerations that are important to erosion control system performance.
4
5 4.3.2.1.3   Integration and Interaction of Materials
6
7 The assessment of the effectiveness of each engineered barrier may need to account for the
8 interaction with other engineered barriers that may be used (e.g., durability of a cement barrier
9 may be affected by corrosion of an exposed steel liner that transects the cement barrier).
10
11 The reviewer should evaluate the compositions of the materials proposed for the engineered
12 barriers, the spatial location and distribution of the materials, and potential interactions both
13 among the engineered materials and with the natural system. For example, the amount of
14 water that penetrates the engineered barriers, the composition of the penetrating water, and the
15 composition of the water after interaction with the engineered barriers and wasteforms will affect
16 the leaching of radionuclides from the wasteforms and the near-field transport of radionuclides.
17
18 4.3.2.1.4   Construction Quality and Testing
19
20 The reviewer should examine the parameters chosen to represent the engineered barriers in the
21 performance assessment and compare those parameters to the quality requirements for the
22 design and test results of the engineered barriers. Selection of deterministic parameters or
23 parameter distributions should account for expected variability in materials, construction
24 implementation, and other uncertainties (e.g., interactions among materials and with the natural
25 system; the properties of as-emplaced materials). The reviewer should evaluate tests or
26 measurements used to support parameter values implemented in the performance assessment
27 (e.g., permeability and hydraulic conductivity testing). The reviewer should evaluate information
28 developed (or the plans to develop information) to demonstrate that as-emplaced properties are
29 consistent with laboratory-measured or design values.
30
31 4.3.2.1.5   Modeling of Engineered Barriers
32
33 The objective of engineered barrier analysis is to establish model representations of the
34 physical dimensions and characteristics of designed engineered features and to determine the
35 ranges of parameter values that would reasonably represent the behavior of the features with
36 the passage of time (NRC, 2000). In developing the performance assessment model for the
37 engineered barriers, DOE should present a design concept that includes information on spatial
38 relationships among physical components (e.g., the layout and physical dimensions of a vault or
39 cover system) and the physical distribution of various types of materials that are used in the
40 facility. Not all design features will necessarily be reflected in the performance assessment as
41 engineered barriers, but DOE should identify and include those components (e.g., engineered
42 caps) and associated materials (e.g., reducing grout) that are most important to demonstrating
43 compliance with the performance objective. The reviewer should examine those DOE-identified
44 components and materials and evaluate how they are represented in the performance
45 assessment modeling of the engineered barriers.
46
47 The reviewer should evaluate the degradation mechanisms associated with the engineered
48 barriers. Barriers may degrade from internal (e.g., interaction between incompatible materials,
49 interaction with the waste) or external processes (e.g., interaction with biota, erosion, leaching
50 by infiltrating water, disruptive processes such as seismically induced cracking). Analysis of a

1 barrier system should be performed in an integrated manner because of the potential synergism
2 between degradation mechanisms. If the analysis is performed assuming the degradation
3 mechanisms are independent, the reviewer should evaluate the information to determine
4 whether adequate basis has been provided for the analysis approach (e.g., assuming the
5 degradation mechanisms can be evaluated as independent), which may include demonstrating
6 that the degradation analysis was reasonably conservative.
7
8 The reviewer should evaluate how DOE has considered the interaction of the components of the
9 engineered system and materials in the engineered barrier system. Factors that may need to
10 be considered include (1) compatibility among materials that may come in contact with each
11 other; (2) the manner in which construction may affect system behavior (e.g., construction joints,
12 changes in geometry, penetrations); (3) the effect that failure of a design feature or some
13 portion of an engineered barrier would have on the overall behavior of the system; and (4) how
14 the degradation of material properties affects barrier performance over time (NRC, 2000). The
15 DOE performance assessment should account for relevant materials and conditions that could
16 affect release from the waste disposal system over the service life of the engineered barriers.
17
18 The reviewer should evaluate information that DOE uses to support the model estimates of
19 engineered barrier performance. This may include site-specific test information, previous
20 experience with similar systems, process modeling of barrier component performance
21 (e.g., detailed models of an infiltration cap), field studies, natural analogs, independent peer
22 review, or additional sources of relevant information. DOE may also use preliminary analyses to
23 assess the need for additional performance enhancements that may, in turn, dictate the use of
24 improved or additional engineered barrier systems (e.g., the performance modeling of reinforced
25 concrete vaults, soil covers). In this manner, design features and the design of engineered
26 barriers would evolve from important conclusions developed with initial performance
27 assessment results. Information from these types of analyses may be factored into monitoring
28 activities (see Section 10).
29
30 **4.3.2.2   Review Procedures**
31
32 To review models of engineered barriers, consider the degree to which DOE relies on
33 engineered barriers and near-field radionuclide transport to demonstrate compliance with
34 10 CFR Part 61, Subpart C, and the contribution of the engineered barriers to system
35 performance (see Section 4.6). For example, if DOE relies on the engineered barriers to
36 significantly reduce the mass flux of waste from the disposal system compared to that provided
37 by the waste and natural system, then perform a detailed review of the modeling. On the other
38 hand, if DOE demonstrates the model to have a minor impact on the dose of the receptor, then
39 conduct a simplified review to determine whether the calculations of barrier performance have
40 been appropriately implemented in the performance assessment. In a risk-informed,
41 performance-based review, some of the review procedures may not be necessary when
42 conducting a simplified review for those models that have a minor impact on performance. The
43 reviewer should perform the following procedures:
44
45 •   Apply the general review procedures found in Section 4.2 to the assessment of
46   engineered barrier performance.
47
48 •   Evaluate the descriptions of the engineered barriers proposed for the site, and determine
49   whether the descriptions adequately describe the physical and chemical characteristics
50   of the barriers.

- Confirm that the design for the engineered barriers adequately provides the dimensions, orientation, and compositions of the barriers. Specifically, the reviewer should examine figures and illustrations (e.g., cross sections that illustrate the components of the engineered barrier system).

- Assess whether the descriptions of the engineered barriers adequately detail the design features, the functionality (e.g., ability to limit water contact with the waste), and properties (e.g., porosity, hydraulic conductivity, sorption coefficient) of the barriers.

- Verify that an adequate description of the materials and methods used to construct the engineered barriers has been provided.

- Ensure that the description of degradation mechanisms and physical and chemical phenomena that may affect the degradation of the engineered barriers is clear and complete and any synergisms between mechanisms are described. For example, degradation mechanisms for a cementitious barrier may include sulfate and magnesium attack, carbonation, reinforcement corrosion, leaching, alkali/aggregate reactions, freeze-thaw cycling, cracking (e.g., thermal, seismic-induced), and shrinkage.

- Examine the assumptions about how the barriers will degrade and the justification and technical bases for the time period for which DOE takes credit for the effectiveness of the barriers.

- Verify that mathematical models for the degradation of engineered barriers and near-field transport of radionuclides are based on similar environmental parameters, material properties, and assumptions.

- Evaluate the technical bases used to support estimates of physical durability of engineered barriers with time. For example, ensure that conceptual models for fracturing of a wasteform or clogging of a drainage layer in an engineered cap are supported by test results that are appropriate for the materials used in the barrier.

- Evaluate the technical bases DOE provided to support estimates of the chemical performance of engineered barriers and wasteforms (e.g., reducing grout, saltstone) during the performance period.

- Ensure that DOE has evaluated potential changes in pore water chemistry (e.g., pH, redox) with time, taking into account the amount of water expected to pass through the wasteform and the proposed grout formulation (e.g., novel ingredients, cement fraction).

- Ensure that DOE has adequately considered the potential for oxidizing conditions in the disposal system (e.g., oxidation by aqueous or gaseous transport into the wasteform or oxidation of engineered materials by interaction with the waste).

- Ensure that DOE has adequately characterized and considered the potential impact of the waste on cementitious material degradation and radionuclide release (e.g., waste components may increase Eh or decrease pH of the grout that is in close proximity to the waste).

- Evaluate the potential for the physicochemical conditions the barriers produced to limit radionuclide mobility (e.g., by maintaining reducing water compositions).

- Evaluate the impacts of biointrusion (e.g., root penetration, burrowing animals) on engineered barrier performance, and if appropriate, verify that the impacts have been appropriately represented in the performance assessment modeling.

- Evaluate features, events, and processes that can affect the ability of the plant community to transpire water for evapotranspiration barriers, including but not limited to variability in weather conditions, biotic processes, discrete pathways, disease, land use, and disruptive events such as fires.

- Ensure the storage capacity of evapotranspiration covers has been designed considering the performance period and requirements for moisture management at the site.

- Evaluate the compositions of the materials proposed for the engineered barriers and potential interactions among the materials and with the natural system. Ensure that interactions that could affect barrier degradation have been adequately represented in the performance assessment.

- Ensure that DOE has adequately characterized and considered the potential impact of the service environment on cementitious material degradation and radionuclide release. Ensure the degradation mechanisms are consistent with site conditions (e.g., biodegradation [may affect CO_2 concentrations]; concentrations of sulfate, magnesium, chloride, oxygen, and CO_2; hydrological cycle and potential for flooding and wet/dry cycling).

- Ensure that DOE has adequately characterized and considered the effects of grout components on cementitious material degradation and radionuclide release (e.g., effect of supplementary grout components on pH and Eh evolution and presence of deleterious species that may enhance cementitious material degradation or radionuclide release).

- Review grout formulations and associated design specifications for consistency with performance assessment assumptions regarding the expected performance of the grouts (e.g., material property assignment [hydraulic conductivity] are consistent with design specifications [water-to-cement ratio]).

- Review grout placement information and controls or procedures used to ensure the quality of as-emplaced grout is consistent with performance assessment assumptions.

- Ensure relevant features, events, and processes affecting cementitious material degradation, including spatial and temporal transients, are adequately considered (e.g., alternating freeze/thaw, wet/dry cycling, flooding, moisture profiles and distribution, climate change, land use).

- Examine the parameters chosen to represent the engineered barriers in the performance assessment. Compare those parameters to the quality requirements for the design and to the results of tests or measurements of the engineered barrier properties.

- Evaluate any testing that supports parameter values used in the performance assessment (e.g., permeability and hydraulic conductivity testing), and determine whether the test conditions were representative of the expected environmental conditions for the barrier. Determine whether test results have been interpreted appropriately (e.g., that leach test results have been corrected for changes in surface area-to-volume ratios).

- Examine the engineered barrier components and materials DOE identified, and evaluate how they are represented in the engineered barrier performance assessment.

- Evaluate the engineered barrier component modeling in the performance assessment. Examine modeling of interactions between materials, construction effects (e.g., joints, penetrations), potential effects of failure of design features, and degradation of material properties over time.

- Evaluate the parameters used to describe flow through and out of the engineered barriers, and confirm that they sufficiently bound the flow through the barriers.

- Evaluate information used to support the engineering barrier analysis. Examine site specific tests, information on previous experience with similar systems, process models of barrier component performance (e.g., detailed modeling of an infiltration cap), natural analogs, independent peer review, or plans to develop additional model support for engineered barrier system performance.

- Ensure uncertainties in cementitious material degradation were adequately considered and propagated through the performance assessment model.

4.3.3   Source-Term/Near-Field Release

The modeling of the source term can be one of the most important determinants of the predicted overall disposal facility performance. The source term is the inventory, physical and chemical characteristics, and other properties of the waste used to estimate release rates. Source-term modeling estimates the partitioning in and release of radionuclides from the disposal unit. Releases generally occur by advective or diffusive mechanisms, although direct release mechanisms may be possible (e.g., biointrusion, erosion). The near field is generally defined as the area surrounding the waste that may have moisture flow and chemical conditions (e.g., due to the presence of the waste or engineered barriers) significantly different from the natural system in which the waste disposal facility is located.

The source-term analysis calculates radionuclide releases from the disposal facility as a function of space and time. These radionuclide release rates can then be used as input for transport models that estimate offsite releases from the facility. Radionuclides are typically released from the waste or wasteform and transported in the aqueous phase, but release of certain radionuclides (e.g., C-14, H-3, Kr-85) can occur in the gaseous phase. Although liquid releases can be significantly constrained by limitation of the flux of water entering a disposal unit, gaseous releases are relatively unconstrained because of the significantly higher rates for gaseous diffusion and advection compared with diffusion and advection of radionuclides in the liquid phase. Gaseous advection and diffusion may become limited at high liquid saturations. Gaseous and liquid releases will often be analyzed separately in performance assessment analyses because of the significant differences in the nature of the releases, and because in

1 many cases, the limited inventory associated with the gaseous release and limited resultant
2 impact on performance readily lends itself to a simple bounding analysis.
3
4 **4.3.3.1   Areas of Review**
5
6 This portion of the review focuses on the assumptions, data, and models (conceptual and
7 computational) DOE used to develop the source term for the performance assessment model.
8 Source-term analyses calculate releases of radionuclides as a function of time and space. The
9 release rates are used as inputs to radionuclide transport models. The complexities of most
10 sites and proposed disposal approaches usually result in source-term analyses being
11 developed on a site-specific basis. Source-term modeling is commonly implemented in
12 performance assessment models with a simplified representation of both the distribution of the
13 radionuclide inventory and the physiochemical processes associated with the partitioning of
14 radionuclides between the materials and physical phases in the disposal unit. The simplified
15 representation is abstracted for inclusion in mathematical and computer model representations
16 of the real system. The source-term model should include the effects of the degradation of the
17 wasteform and the engineered barriers, as appropriate. For example, cracking of a grouted
18 wasteform over time may lead to advective release, rather than diffusive release. In another
19 example, chemical barriers, such as reducing grout formulations, may lose their effectiveness
20 over time. The release models used in the source-term analysis should reflect the temporal
21 changes to the disposal facility.
22
23 Representing the source term in a performance assessment involves generalizing system
24 details into more simplified conceptualizations that can be modeled. Whereas the source-term
25 abstractions must adequately represent the features and processes significant to disposal
26 system performance, the abstractions must not simplify system behavior to the extent that
27 disposal system performance is significantly underestimated or unrealistically overestimated.
28
29 4.3.3.1.1   Inventory of Radionuclides in Waste
30
31 The inventory of radionuclides in the waste is used to assess the removal of highly radioactive
32 radionuclides to the maximum extent practical (see Section 3) and the concentration limit criteria
33 related to 10 CFR 61.55 (whether the waste exceeds the concentration limits for Class C waste)
34 (see Section 3.5). The inventory also provides the radionuclide inventory for which release
35 rates are estimated with source-term calculations. The radionuclide inventory evaluated in this
36 portion of the review should be consistent with radionuclide inventory used to assess
37 compliance with the site-specific radionuclide removal and concentration limit criteria described
38 in Section 3 of this guidance document.
39
40 The reviewer should evaluate the description of the radionuclide inventory in the waste. All
41 radionuclides (particularly highly radioactive radionuclides) should be described by volume,
42 concentration, and location within the disposal system. Radionuclides with relatively high
43 solubility, low sorption, and high dose conversion factors and/or significant ingrowth are of
44 particular significance. Additional detailed guidance for reviewing radionuclide inventory is
45 provided in Section 3.1.
46

1   <u>4.3.3.1.2   Degradation and Release From Wasteforms</u>
2
3   The reviewer should evaluate the descriptions of the wasteforms and the representation of the
4   wasteforms in the source-term modeling. Waste is almost exclusively stabilized in a solid form
5   to reduce its mobility and dispersibility into the environment. Wasteforms can include
6   cement-solidified waste, activated metal, glass, bulk waste, and others. Wasteforms limit
7   aqueous and gaseous releases once the engineered barriers degrade. Different wasteforms
8   will have different release processes and degradation mechanisms. For example, a high-quality
9   cement-solidified wasteform may be dominated by diffusional release, whereas a glass
10 wasteform may release radionuclides mainly by dissolution of the glass matrix. In addition, the
11 release mechanisms may change during the period of performance (e.g., while diffusion may
12 dominate release from high-quality, intact cementitious wasteforms, advection may dominate
13 release from degraded or lower quality cementitious wasteforms). These differences can be
14 very important in evaluating the appropriateness of source-term models.
15
16 The reviewer should evaluate the wasteform degradation processes DOE considered and
17 evaluate how the processes are incorporated in the source-term model. Wasteform degradation
18 modes may include leaching, dissolution, and other chemical reactions with groundwaters.
19 Cementitious wasteform performance may be reduced by chemical processes such as sulfate
20 attack or carbonation or by physical processes such as cracking caused by settling or seismic
21 activity and damage by reinforcement corrosion. Degradation of the wasteform can increase
22 contact of water with the waste and can provide shortened and more permeable paths for
23 radionuclide release. Degradation can also change the type of mechanisms that dominate
24 release from the wasteform (e.g., cracking may enhance advective release).
25
26 <u>4.3.3.1.3   Source-Term Models</u>
27
28 The reviewer should evaluate the source-term models used. Source-term models are ultimately
29 used to estimate release rates from the disposal facility; many intermediate processes may be
30 simulated to estimate release rates. Release rates can be affected by the performance of
31 engineered barriers, as well as the specific physical and chemical properties of the disposal
32 system and the interaction of the disposal system with the natural environment (e.g., infiltrating
33 water that is high in magnesium may have an effect on cement performance). Some disposal
34 plans will require detailed consideration of these processes and conditions, whereas simplified
35 analyses may be justified for other sites and disposal options. Sites for which the source term
36 models need to be considered carefully are sites for which there is significant credit taken for
37 some aspect of the source-term modeling (i.e., low solubility limits) and for which there is limited
38 model support. Sites for which a simpler analysis is acceptable are sites for which the simple
39 analysis can be shown to be clearly conservative or for which the simple model is well
40 supported by multiple lines of evidence, including field tests that show the simple model
41 accurately represents or bounds field results. The information reviewed as part of the
42 evaluation performed for Section 4.6 should help focus the reviewer on the key aspects of the
43 disposal facility performance and should provide information to determine the importance of
44 source-term modeling and near-field release.
45
46 There are generally four types of aqueous radionuclide releases: (1) rinse release,
47 (2) diffusional release, (3) dissolutional release, and (4) partitioning release. Rinse release
48 refers to washing of radionuclides from the surface of a wasteform by infiltrating groundwater.
49 Diffusional releases occur when radionuclide movement through a porous wasteform (e.g., a
50 cement-stabilized wasteform) is limited by diffusion. Radionuclide releases resulting from

1 corrosion of an activated metal or dissolution of glass wasteforms are examples of dissolutional
2 release. Partitioning release results when radionuclide release is described by a characteristic
3 K_d or other parameter that distributes activity between phases in the system (e.g., between the
4 wasteform and liquid contacting the wasteform). Solubility limits may be very important in
5 estimating release rates with source-term models, particularly for extreme chemical
6 environments (e.g., high pH associated with pore fluids of cementitious wasteforms or
7 cementitious engineered barriers).

9 ### 4.3.3.1.4   Chemical Environment

11 The reviewer should evaluate the chemical environments of the system. There may be spatial
12 variation in the chemical environments within the system, and they may also change over time.
13 The reviewer should evaluate the chemical environment for consistency with the degradation
14 of wasteforms and engineered barriers that may affect the chemical environment (see
15 Section 4.3.3.1.2). The chemical environment may be important for estimating the lifetime of
16 engineered barriers. The chemical environment is important to estimating radionuclide release
17 from the wasteform and transporting released radionuclides within and from the disposal facility.
18 The chemical environment is particularly important if DOE relies on solubility limits or retardation
19 of radionuclides within the disposal unit to satisfy the performance criteria. The site-specific
20 chemical environment of a disposal facility may include engineered components designed to
21 have performance-enhancing chemical properties.

23 ### 4.3.3.1.5   Gaseous Releases

25 The reviewer should evaluate the potential for gaseous releases because some radionuclides
26 may be released in a gaseous form (e.g., C-14, tritium). The timing and rate of a gaseous
27 release will depend on the engineered barrier design. After release from the engineered
28 barriers, gaseous radionuclides can move by advection and diffusion through overlying soil or
29 other materials to reach the atmosphere. Some gaseous releases may interact with
30 components of the soil and groundwater and are not strictly controlled by advection and
31 diffusion. Gaseous releases are likely to have limited impact on most waste determinations due
32 to limited inventory, and they may be handled with relatively simple bounding calculations
33 (e.g., a box model). The reviewer should evaluate assumptions regarding the effects of
34 saturation on diffusivities and chemical and biological retention.

36 **4.3.3.2   Review Procedures**

38 Review the source-term modeling considering its importance to disposal facility performance as
39 evaluated in accordance with Section 4.6. If DOE relies on the source term to significantly
40 reduce or mitigate radiological impacts, then perform a detailed review of the source-term
41 modeling. If, on the other hand, DOE demonstrates that this abstraction has a minor impact on
42 the dose, then conduct a simplified review focusing on determining whether the calculations
43 have been properly implemented in the performance assessment model. The reviewer should
44 evaluate the consistency of source-term information with information provided on engineered
45 barrier systems (Section 4.3.2) and on climate and infiltration (Section 4.3.1). The reviewer
46 should verify that source-term modeling has been appropriately integrated with other models in
47 the performance assessment. In a risk-informed, performance-based review, some of the

review procedures may not be necessary when conducting a simplified review for those models that have a minor impact on performance. The reviewer should perform the following procedures:

- Apply the general review procedures found in Section 4.2 to source-term and near-field release modeling.

- Evaluate the description of the radionuclide inventory in the waste. Confirm that radionuclides are described by volume, concentration, and location within the disposal system. Information evaluated should be consistent with that considered under Section 3.1 of this guidance document.

- Examine the description of the wasteform, and verify that the implementation of the wasteform in the source-term modeling for the performance assessment is consistent with the description.

- Examine the DOE description of environmental conditions expected inside failed engineered barriers and within the disposal facility environment surrounding the engineered barriers. Verify that the ranges in conditions are described in sufficient detail and ranges of chemical conditions have been appropriately accounted for in the performance assessment.

- Evaluate potential changes to the chemical environment of the disposal system over time and resulting changes in degradation of the engineered barriers and wasteforms that may affect the source term and near-field transport. Verify that these potential changes are consistent with the engineered barrier performance information in Section 4.3.2 of this guidance document. Verify that the uncertainty in changes in the chemical conditions has been propagated in the performance assessment, if needed.

- Evaluate radionuclide release rate test programs and other sources of data supporting the durability of and release rates from the wasteforms. Verify that the programs or data sources provide sufficient and suitable data for use in the source-term abstraction. Evaluate the justification for the use of test results not specifically collected from the site of interest.

- Evaluate the parameters used to describe flow through and out of the wasteform, and confirm that they sufficiently bound the flow through the wasteform.

- Evaluate the potential for gaseous releases of radionuclides. Verify that potential gaseous releases are consistent with the design of the engineered barriers evaluated in Section 4.3.2.

- Evaluate assumptions regarding the effects of saturation on diffusivities and chemical-and biological-mediated attenuation of potential gaseous releases.

- Verify that DOE has adequately considered the uncertainties in the characteristics of the natural system and engineered materials (e.g., the type, quantity, and reactivity of material) in establishing initial and boundary conditions for conceptual models and simulations of processes that affect the source term.

- Confirm that DOE has considered a range of wasteform degradation mechanisms that are appropriate to the wasteform design and the physical and chemical conditions of the disposal environment.

- Confirm that data used to support the release rates were developed for materials and conditions that encompassed the range of composition of the wasteform and the range of chemical conditions of the disposal environment.

- Ensure that changes in release mechanisms that could occur because of degradation of the engineered barriers are appropriately accounted for (e.g., advective versus diffusive release from degraded wasteforms).

- Ensure that moisture characteristic curve parameters used for near-field release modeling are supported by empirical measurement or, if generic information is used, the generic information is sufficiently conservative.

- Ensure that selection of K_d and solubility limit parameters used in the release model adequately considered the material and chemical environment of the wasteform (e.g., literature values relevant to ordinary cement may not be relevant to novel grout formulations). Ensure that the effects of additives to the wasteform or barrier and the presence of chelating agents in the waste have been considered in developing sorption coefficients and solubility limits.

- Ensure that changes in the chemical condition of the wasteform with time (see Section 4.3.3.1.2) are appropriately reflected in the release model (e.g., changes in K_d and solubility limits with pH and redox conditions).

- If radionuclide release is based on linear partitioning, confirm that uncertainty in the K_d values is reflected in the release model. In general, literature-based values are more uncertain than site-specific values.

- Evaluate whether a solubility or partitioning release model is appropriate, and ensure that it is representative or conservative.

- Ensure that the release model is calibrated to release test data, scaling for surface area/volume, or that plans have been developed to acquire release rate data from surrogate or actual wasteforms.

4.3.4 Radionuclide Transport

4.3.4.1 Areas of Review

This section focuses on evaluating transport of radionuclides beyond the engineered barriers. Transport of radionuclides to the receptor group(s) may be through air, water, or biotic pathways. The reviewer should consider the dimensions, locations, and spatial variability of the various transport pathways as well as temporal variations during the compliance period. The information reviewed for radionuclide transport should be consistent with the general information evaluated using Section 1 of this guidance document.

1 ### 4.3.4.1.1   Air Transport
2
3 The reviewer should evaluate the potential for airborne transport of radionuclides.
4 Reviewers evaluating airborne transport processes should consider both suspension of
5 radionuclide-bearing particulates and release of gaseous phase radionuclides (e.g., H-3, C-14,
6 Kr-85). The reviewer should evaluate the significance of dilution and dispersion of the airborne
7 radionuclides as they are transported in the atmosphere. Atmospheric transport of gaseous
8 radionuclides will be affected by the height of the release above ground level, the speed and
9 direction of the wind, atmospheric stability, and terrain. Radionuclide transport in the air will be
10 affected by rainfall and particulate settling.
11
12 Information DOE provided on airborne transport should be consistent with the site description
13 (e.g., meteorology) (Section 1.1.3), information on inadvertent intrusion (e.g., farming or drilling)
14 (Section 5), and climate (Section 4.3.1).
15
16 ### 4.3.4.1.2   Surface Water Transport
17
18 The reviewer should evaluate information provided on the potential for radionuclides to be
19 transported beyond the engineered barriers of the disposal facility via surface water. In most
20 cases, radionuclides will be transported from the waste disposal facility via other pathways
21 before being transported in surface water pathways because it is unlikely surface water will
22 directly intersect the waste disposal facility. Mechanisms for radionuclides to enter surface
23 waters include but are not limited to deposition after airborne transport, groundwater discharge,
24 and overland flow (e.g., associated with erosion). Information DOE provided on surface water
25 transport should be consistent with the information provided on climate and infiltration
26 (Section 4.3.1) and with the groundwater transport analyses for the site. The reviewer should
27 evaluate information provided on potential dilution of radionuclide concentrations by mixing of
28 disposal facility releases with surface waters. Typically, transport and residence times in
29 surface water systems are relatively short; therefore, dispersion and dilution are the dominant
30 processes that will mitigate the impact of contaminants released to most surface water bodies
31 such as streams and rivers. In addition to transport in the water itself, the reviewer should
32 consider the potential for radionuclide transport along with sediment suspended in the
33 surface water.
34
35 The reviewer should evaluate the chemistry of the surface water and host rocks and sediments
36 with respect to the potential for transport of radionuclides. The reviewer should evaluate the
37 effects of speciation of radionuclides (e.g., due to sorption, precipitation, ion exchange) in the
38 surface water if performance assessment modeling accounts for these processes. The reviewer
39 should evaluate information provided on flooding at the site (e.g., flood hydrographs, probable
40 maximum floods, maps of drainage basins, and maps of floodplains) (see Section 7).
41
42 ### 4.3.4.1.3   Transport in the Unsaturated Zone
43
44 The reviewer should evaluate the information provided on the potential for radionuclides to be
45 transported beyond the engineered barriers of the site along groundwater pathways through the
46 unsaturated zone to the water table. Groundwater transport is among the most likely
47 processes for radionuclides to be transported from the engineered systems of the disposal
48 facility. The reviewer should examine the hydrogeologic data for the site, including the
49 stratigraphy and geologic structures (e.g., fractures) that may affect groundwater flow,
50 thicknesses of unsaturated strata, unsaturated hydraulic properties, depth to groundwater

1   (including any perched zones that may affect transport), and recharge to and discharge from the
2   site (including the potential effects of climate change). The reviewer should evaluate the
3   information provided on the potential for diffusion and mechanical dispersion during transport.
4   The reviewer should examine the significance of spatial variations in hydrogeologic properties
5   and examine the site design for information showing the positions of the engineered structures
6   and anthropogenic features (e.g., infiltration caps) that may influence the unsaturated zone
7   hydrology of the site. The reviewer also should evaluate the unsaturated zone flow and
8   transport models DOE used in its performance assessment and how output is passed from the
9   unsaturated zone flow and transport models to the saturated zone flow and transport models.
10 The staff should review the lengths of the flow paths in the unsaturated zone and evaluate the
11 travel times estimated in the analysis. Information DOE provided on groundwater transport
12 should be consistent with the information provided for the site description (Section 1.1.3),
13 climate and infiltration (Section 4.3.1), and surface water transport (Section 4.3.4.1.2).
14
15 The reviewer should evaluate the chemistry of the groundwater and host rocks and sediments
16 with respect to potential radionuclide transport, including potential changes to the chemistry of
17 the groundwater arising from interactions with the disposal system components and wasteform.
18 The reviewer should evaluate the effects of speciation of radionuclides in the groundwater
19 (e.g., due to sorption, precipitation, ion exchange, redox reactions) and the potential for colloid
20 facilitated transport. The information DOE provided should be consistent with the information
21 provided for the radionuclide inventory (Section 3.1), engineered barriers (Section 4.3.2),
22 source-term models (Section 4.3.3), and chemical environment (Section 4.3.3.1.4).
23
24 4.3.4.1.4   Transport in the Saturated Zone
25
26 The reviewer should evaluate the information provided on the potential for radionuclides to be
27 transported from the disposal facility along groundwater pathways in the saturated zone to
28 receptors. The reviewer should examine the hydrogeologic data for the site, including
29 information provided describing the aquifers, aquitards, and geologic features (e.g., fractures)
30 that may affect groundwater flow in the saturated zone. The reviewer should examine the
31 significance of spatial variations in hydrogeologic properties. The reviewer should evaluate
32 DOE estimates of recharge to and discharge from the aquifers; groundwater flow velocities,
33 gradients, and volumes; and ambient groundwater compositions. Water withdrawals and
34 pumping of saturated zone aquifers may affect flow, especially dilution of radionuclide
35 concentrations at the withdrawal point. The reviewer should evaluate information provided on
36 the potential for diffusion, mechanical dispersion, decay, and ingrowth during transport. The
37 reviewer should examine information provided on engineered structures (e.g., slurry walls) that
38 may affect saturated zone flow and transport at the site. The reviewer should evaluate the
39 saturated zone flow and transport models DOE used in the performance assessment. The staff
40 should review the lengths of the flow paths in the saturated zone and evaluate the travel times
41 estimated in the analysis. Information DOE provided on saturated zone groundwater transport
42 should be consistent with the information provided for the site description (Section 1.1.3),
43 climate and infiltration (Section 4.3.1), surface water transport (Section 4.3.4.1.2), and
44 groundwater transport in the unsaturated zone (Section 4.3.4.1.3).
45
46 The reviewer should evaluate the chemistry of the groundwater and host rocks and sediments,
47 including potential changes to the chemistry of the groundwater arising from interactions with
48 the disposal system components and wasteform. The reviewer should evaluate the effects of
49 speciation of radionuclides in the groundwater (e.g., due to sorption, precipitation, ion
50 exchange, redox reactions) and the potential for colloid facilitated transport. The information

1 DOE provided should be consistent with the information provided for the radionuclide inventory
2 (Section 3.1), engineered barriers (Section 4.3.2), source-term models (Section 4.3.3), and
3 chemical environment (Section 4.3.3.1.4).
4
5 **4.3.4.2   Review Procedures**
6
7 Review the transport modeling considering the importance of this information to the
8 performance of the disposal facility as evaluated using the review procedures found in Section
9 4.6. If DOE relies on the transport modeling to significantly reduce or mitigate radiological
10 impacts, then the reviewer should perform a detailed review of the transport modeling. For
11 example, if DOE relies on retardation during transport to significantly delay the transport of
12 radionuclides, then the reviewer should perform a detailed review of the transport modeling. On
13 the other hand, if DOE demonstrates the transport modeling to have a minor impact on the
14 estimated radiological dose to the receptor, then the reviewer should conduct a simplified
15 review. In a risk-informed, performance-based review, some of the review procedures may not
16 be necessary to review models with a minor impact on predicted performance. The reviewer
17 should perform the following procedures:
18
19 •   Apply the general review procedures found in Section 4.2 to the modeling of
20   radionuclide transport.
21
22 •   Evaluate the potential for airborne transport of radionuclides. Verify that both
23   suspension of radionuclide-bearing particulates and release of gaseous phase
24   radionuclides are adequately considered. Examine the potential significance of dilution
25   and dispersion along the atmospheric transport path.
26
27 •   Confirm that DOE has developed adequate technical basis for the dilution and
28   dispersion of radionuclides by mixing in surface waters during transport. Evaluate
29   whether adequate technical basis has been provided for incorporating the impact of
30   variability, especially temporal, in surface water dilution.
31
32 •   Determine whether the development of models and data to represent the sorption and
33   speciation of radionuclides during surface water transport accounted for the chemistry of
34   the surface water and the mineralogy of sediments.
35
36 •   Confirm that DOE has considered the potential for radionuclide transport along with
37   sediment moved by the surface water, in addition to transport in the water itself. Verify
38   that the models are consistent with information provided on flooding potential at the site.
39   For example, ensure that the potential for erosion and exhumation of wasteforms is
40   considered for those areas where site characteristics (e.g., gullies, steep terrain) indicate
41   that erosion is a significant process (see Section 7).
42
43 •   Verify that the description of the hydrology, geology, climatology, geochemistry, design
44   features, and physical and chemical phenomena that may affect radionuclide transport
45   is adequate.
46
47 •   Verify that the assumptions in radionuclide transport modeling are clearly identified and
48   are consistent with the relevant data presented in the system description.
49

- Verify that DOE has adequately described groundwater flow directions and velocities (horizontal and vertical) for each potentially affected aquifer. When applicable, the groundwater hydrology should be described by making use of hydrogeologic columns, cross sections, and water table and/or potentiometric surface maps.

- Verify that the information on groundwater flow direction in each hydrological unit is consistent with the information presented about receptor location reviewed in Sections 1.1.3.1 and 4.1.1.4. Ensure that the groundwater flow directions are consistent with placement of a member of the public at the point of highest exposure outside of the disposal area to demonstrate compliance with 10 CFR 61.41.

- Confirm that DOE has sufficiently described numerical analysis techniques used to characterize the unsaturated and saturated zones, including the model type, justification, documentation, verification, calibration, and other associated information. In addition, verify that DOE adequately described the input data, data generation or reduction techniques, and any modifications to these data.

- Evaluate the adequacy of the description of the effects of radionuclide speciation on processes that could affect radionuclide transport in groundwater (e.g., sorption, precipitation, ion exchange, redox reactions) and the potential for colloid-facilitated transport.

- Ensure that the selection of transport properties (e.g., K_d values) used in the unsaturated and saturated zone transport models adequately considered the mineralogy and water chemistry of the system.

- Evaluate the technical basis for the transport parameters and determine whether DOE modeling assumptions for radionuclide transport are appropriate. For example, confirm that the selected geochemical parameters (e.g., pH, redox, sorption coefficients) are consistent with the expected chemical environment at the site.

- Evaluate whether chelating agents or other organic solvents have been used at the site and released to the environment. If chelating agents have been released, evaluate how they have been considered in selecting transport properties and whether the model support adequately constrains the impact of chelating agents on radionuclide transport. Evaluate the technical basis on whether the impact of chelating agents on transport properties is expected to persist when radionuclides may eventually be released.

- Confirm that DOE has used flow and transport parameters that are based on techniques that may include laboratory experiments, field measurements, information from comparable sites, and process-level modeling studies conducted under relevant conditions. Confirm that site-specific information was used, when available, to develop transport parameters and models.

- Ensure that adequate descriptions are provided of how flow and transport data were used, interpreted, and incorporated into the performance assessment parameters.

- Ensure that DOE provided hydrologic properties (e.g., moisture characteristic curve parameters) for modeling unsaturated zone flow that are supported by empirical measurement. If generic information is used, ensure that the properties are sufficiently conservative. Verify that the information is consistent with related hydrogeologic data for the site, including observed spatial variability in the hydrologic properties.

- Verify that parameter values for processes such as matrix diffusion, dispersion, and groundwater mixing are based on reasonable assumptions about climate, aquifer properties, and groundwater volumetric fluxes.

- Confirm that the uncertainties in transport properties (e.g., K_d values) are reflected in the unsaturated and saturated zone transport models. In general, literature-based values are more uncertain than site-specific values.

- Ensure that limitations and uncertainties of the K_d model have been adequately considered if a K_d model is used to represent radionuclide transport and if the transport submodel has been identified as having a significant effect on dose results (e.g., by significantly delaying radionuclide arrival and resulting in significant decay). The reviewer may consider "Understanding Variation in Partition Coefficient, K_d, Values" (EPA, 1999).

- Examine the results of DOE field transport tests or observations of leaks and spills, and verify that the performance assessment model results are consistent with the field experiments or observations, or confirm that an adequate technical basis has been provided to explain any significant differences. For example, the performance assessment transport models should provide results that are consistent with the transport of existing contaminant plumes as observed through environmental monitoring.

- Ensure that the output from the unsaturated zone flow and transport model is consistent with the input into the saturated zone flow and transport model.

- If the water table is shallow and the unsaturated zone flow paths are short {less than 5 m [16 ft]}, ensure that DOE has evaluated the impact of water table fluctuation and provided adequate technical basis that fast pathways that would significantly affect the travel time through the unsaturated zone are not present .

- Ensure that adequate technical basis is provided represent the ingrowth of daughter radionuclides in the modeling of radionuclide transport. Many computer codes assign the same transport properties to the daughter radionuclides that ingrow from parents during transport. If significant ingrowth can occur, the transport of daughters that are more mobile than their parent radionuclide may be significantly underestimated.

- For those radionuclides for which transport in groundwater is important to estimating the dose to receptors, verify that adequate model support has been provided for the transport modeling such as comparison to laboratory experiments, field measurements, observations of leaks and spills, process-level modeling studies conducted under relevant conditions, natural analogs, and independent peer review. For observational types of information such as monitoring of leaks and spills, confirm that the conditions of the leak or spill are similar to the modeled conditions (e.g., mineralogy, pore water chemistry, hydraulic properties).

- Ensure the modeling of saturated transport is consistent with site-specific information about the hydrological units (e.g., information reviewed according to Section 1.1.3.5).

1   Ensure that the modeling of any units identified as aquitards is consistent with
2   information about the spatial variability of the unit. The reviewer should determine that
3   the basis for the absence of any fast pathways through the unit (e.g., areas where the
4   unit pinches out or becomes thin or has high-permeability features such as sand lenses)
5   is adequate. In some cases, it may be conservative to assume an aquitard is intact. As
6   necessary, the reviewer should consider the resolution of the characterization
7   techniques used to identify fast pathways in aquitards.

8

9 •   To the extent practical, consider other relevant sources of information, such as
10   characterization and modeling performed for existing contamination or other waste
11   disposal facilities at the DOE site, that may support or refute the hydrogeologic
12   conceptual model, analysis, and modeling DOE provided for the waste determination.

13

14 •   Ensure that alternative modeling approaches, if applied, are consistent with available
15   data and current scientific knowledge.

16

17 •   Confirm that outputs of radionuclide transport models used in the performance
18   assessment reasonably agree with or bound the results of corresponding process-level
19   models, empirical observations, or both.

20

21 •   Verify that procedures to construct and test the mathematical and numerical models of
22   radionuclide transport are documented and that the procedures are based on modeling
23   approaches that have been accepted by the scientific community.

24

25 **4.3.5   Biosphere Characteristics and Dose Assessment**

26

27 For the purpose of this review, the biosphere is the physical environment accessed by the
28 receptor in the dose assessment. The dose assessment is that portion of the performance
29 assessment model that calculates dose to the receptor from radionuclides transported from the
30 disposal site to the biosphere. The dose assessment includes all the applicable local fate and
31 transport pathways within the biosphere that culminate in exposure to the receptor
32 (e.g., irrigation of soils with contaminated groundwater, plant and animal uptake, consumption
33 of local food products). Because offsite receptor locations can have different characteristics
34 from the disposal site and some specialized information may be needed to support the dose
35 assessment, the biosphere characteristics must be reviewed to verify that the inputs and
36 assumptions have adequate technical basis.

37

38 **4.3.5.1   Areas of Review**

39

40 Input parameters for dose assessments may be generally classified as behavioral, metabolic, or
41 physical. Behavioral parameters collectively describe the behavior hypothesized for the
42 potentially exposed individual, and the behavior is normally consistent with local practices
43 (e.g., time spent gardening, vegetable consumption rates). Metabolic parameters also describe
44 the exposed individual, but generally address involuntary physiological characteristics of the
45 individual (e.g., breathing rates, factors converting intake of unit activity to dose [by
46 radionuclide]). Physical parameters collectively describe the physical characteristics of the site
47 (e.g., geologic, hydrologic, geochemical, ecological, and meteorologic inputs). This section
48 focuses on the review of the physical, behavioral, and metabolic input parameters used in the
49 dose assessment modeling for the performance assessment.

1 4.3.5.1.1   Exposure Pathways and Dose Modeling
2
3 The reviewer should evaluate information provided to ensure that models of radionuclide
4 transport via groundwater, surface water, and air pathways are properly integrated with dose
5 assessment models. For example, groundwater may be used as a source of drinking water
6 (both human and livestock), to irrigate crops, and as a primary source of water for meeting
7 various domestic, commercial, and industrial needs. Therefore, scenarios involving
8 radionuclides transported in groundwater may involve dose modeling that includes direct human
9 ingestion of contaminated water and transfer of contaminants to crops and livestock.
10 Furthermore, partitioning of contaminants to soils can result in inhalation of radionuclides either
11 as particulates suspended in air or as gases emanated from the soils. Direct exposure to
12 contaminated soils is also a potentially applicable dose pathway. Surface water transport can
13 lead to exposure pathways similar to groundwater if the surface water is used as a primary
14 source of water for municipal needs. Direct exposures can occur from contaminated surface
15 water during recreational activities such as bathing or swimming. Air transport of gaseous or
16 particulate releases can result in direct inhalation dose from breathing the air or in applicable
17 soil-related pathway exposures from deposition of particulates to the ground surface. The
18 reviewer should ensure the selected exposure pathways are reasonably complete (e.g., they
19 represent the primary means by which humans can be exposed to radionuclides released to air,
20 groundwater, and surface water) and are consistent with regional practices in the vicinity of the
21 disposal site.
22
23 Justification should be provided for excluding applicable exposure pathways or implementing
24 unique or novel approaches to modeling.
25
26 4.3.5.1.2   Site-Specific Input Parameters
27
28 The reviewer should evaluate site-specific and generic information provided for the biosphere
29 characteristics and dose assessment. When deciding whether site-specific or generic
30 information for biosphere model input parameters should be used, the reviewer would generally
31 consider the characteristics of the scenario and potential receptor groups, the modeled system,
32 what the parameters represent, and how the parameters are used in the code. Based on those
33 considerations, an input parameter value is developed that is appropriate for both the system
34 being modeled and for the conceptual and numerical models implemented by the code.
35
36 Because there is uncertainty associated with the behavior of a hypothetical receptor, it is
37 necessary to rely on generically defined receptors for behavioral and metabolic input
38 parameters. Behavioral and metabolic characteristics of receptors must be representative of
39 average members of the receptor group assumed in the modeled exposure scenario. Some
40 performance assessment codes may use default model values for the behavioral and metabolic
41 parameters. The reviewer should ensure that the use of default parameters is consistent with
42 characteristics described for the average member of the receptor group (e.g., if the average
43 member of the critical group is an adult, then it would be inappropriate to use default soil
44 ingestion rates that are appropriate for a child). Commonly, the average member of the critical
45 group is defined as an adult because adults are exposed to more pathways. However, in
46 certain scenarios with limited pathways, children may be the critical group.
47

1 **4.3.5.2   Review Procedures**

3 The following review procedures focus on evaluating the behavioral and metabolic input
4 parameters used in dose assessments for demonstrating protection of the general population
5 from releases of radioactivity. These review procedures also apply to behavioral and metabolic
6 input parameters used in dose assessments for inadvertent intruder scenarios and scenarios to
7 demonstrate protection of the public during operations. The reviewer should perform the
8 following procedures:

10 •   Examine the coupling of groundwater, surface water, and air transport models to
11   biosphere models. The transport model outputs (e.g., fluxes or concentrations at the
12   biosphere interface) should be linked or have information transferred to the applicable
13   pathways in the biosphere dose assessment model(s).

15 •   Verify that conceptual models for the biosphere include consistent and defensible
16   assumptions based on regional practices and characteristics (i.e., conditions known to
17   exist or expected to exist at the site or surrounding region).

19 •   Confirm that dose assessment results are provided for each exposure pathway
20   (e.g., drinking water, crops, meat, milk, fish, inhalation, external) to assess the
21   reasonableness of pathway contributions to the total dose.

23 •   Verify input parameters and technical bases for the parameters (e.g., transfer factors,
24   consumption rates) for any pathways that are key contributors to dose or have an
25   unexpectedly high or low contribution to the calculated dose.

27 •   Verify the receptor evaluated in the dose assessment has been defined as the average
28   member of the critical group, or a more limiting definition is provided. If the average
29   member of the critical group is selected to be an adult, confirm that the adult is the
30   limiting receptor definition.

32 •   Verify that the internal and external dosimetry approach is consistent with NRC-accepted
33   dosimetry methods (e.g., ICRP 26, ICRP 72) (see Section 4.6.1.3).

35 •   Ensure that selection of the appropriate lung clearance class for inhalation dose
36   coefficients and the fractional uptake to blood for ingestion dose coefficients has an
37   adequate technical basis (e.g., based on the chemical form of the material inhaled or
38   ingested material) or that the highest (most conservative) values of available coefficients
39   are used.

41 **4.4   Computational Models and Computer Codes**

43 **4.4.1   Areas of Review**

45 This section focuses on ensuring that codes used to develop a performance assessment model
46 are appropriately chosen, have undergone quality assurance testing (see Section 8), and that
47 model scenarios and conceptual models reviewed as described in Section 4.1 of this guidance
48 document have been appropriately incorporated into the computational model. The reviewer

1   will benefit from familiarity with NUREG–1757 (NRC, 2003, Vol. 2, Appendix I), which
2   provides guidance for selecting computer codes and incorporating conceptual models into
3   computational models.
4
5   The reviewer should evaluate information provided to ensure that acceptable quality assurance
6   code testing has been conducted (see Section 8). The reviewer is expected to conduct a more
7   detailed and thorough review of less common codes and codes that may have been developed
8   for site-specific application. For example, to complete a review of a well-established
9   commercial product, it may only be necessary to review input files (e.g., for errors such as unit
10 conversion problems) and output files and to ensure the model has been applied over a range
11 of conditions for which the software has been validated. On the other hand, a code developed
12 by DOE or its contractors for a site-specific evaluation may require a more thorough
13 examination of the quality assurance documentation to ensure an appropriate and accurate
14 implementation of the conceptual model.
15
16 The process of developing a performance assessment model typically has many steps. A
17 number of these steps are reflected in associated review procedures in this section instead of in
18 Section 8, because the analysis steps for the performance assessment may not be explicitly
19 represented in quality assurance procedures (e.g., for data or software) but can be important to
20 performance assessment analysis.
21
22 **4.4.1.1   Modeling Approach: Probabilistic or Deterministic**
23
24 DOE may select either a deterministic approach or a probabilistic approach for the analysis to
25 demonstrate compliance with the performance objectives in 10 CFR Part 61, Subpart C. A
26 deterministic analysis uses single parameter values for every variable in the code. By contrast,
27 a probabilistic approach assigns parameter ranges to certain variables, and the code samples
28 and selects the values for each variable from the parameter probability distribution each time
29 the dose is calculated. While a deterministic analysis calculates the results from a single
30 solution of the equations each time the user runs the code, a probabilistic analysis calculates
31 hundreds of solutions to the equations using different values for the parameters from the
32 parameter ranges. The deterministic model, without additional sensitivity analyses, gives no
33 indication of the sensitivity of the results to certain parameters or of the importance of the
34 uncertainty in the parameters. Therefore, applying a deterministic approach may result in the
35 need for stronger justification of code input parameter values and may require further analysis of
36 doses using upper or lower bounding conditions to gain insights on the range of dose estimates.
37
38 A *preferred* method for complex assessments is to use a risk-informed approach to performance
39 assessment. This method uses probabilistic sampling for model parameters with irreducible
40 uncertainty that cannot otherwise be shown to be unimportant to system performance. In this
41 type of model, parameters that are well constrained or that can logically be shown to be of little
42 significance are assigned deterministic values, while the remaining parameters are assigned
43 probability distributions that cover the expected ranges of the parameters. Probabilistic
44 approaches to performance assessment are preferred in some situations because they readily
45 permit propagating and assessing the impact of uncertainty on the model results.
46
47 Although probabilistic approaches are preferred for complex assessments, it is acceptable to
48 use a deterministic model to demonstrate compliance with performance objectives.
49 Performance assessments may be used for many different types of problems, ranging from
50 simple to complex. A simple approach is preferable for a simple problem. This guidance

1   document provides a review framework for different approaches to performance assessment.
2   The choice of the assessment approach is also related to the model support. When the model
3   support is extensive, the uncertainty can be constrained, thereby reducing the importance of a
4   comprehensive uncertainty analysis.
5
6   In general, if deterministic modeling is used, it should be reasonably conservative and
7   sufficiently documented so that a subject matter expert, with minimal interaction with those who
8   performed the modeling, could conclude that the analysis was conservative. Additional
9   sensitivity analyses to identify significant parameters, model components, and processes may
10   be needed if deterministic analyses are performed. Deterministic sensitivity analyses can be a
11   challenge to plan, execute, and evaluate. Nonlinearity in the model (dose) response to
12   independent parameter values can make selection of appropriate parameter sets for sensitivity
13   cases difficult. In addition, local minima or maxima in the response can occur and the
14   parameter space can be large. Review procedures are provided in Section 4.5.2 to facilitate the
15   review of deterministic sensitivity analyses.
16

17   **4.4.1.2   Model Development**

18
19   Sections 4.1–4.3 of this document contain guidance to ensure that the models used in the DOE
20   performance assessment (e.g., source term, water infiltration, engineered barrier performance,
21   radionuclide transport, receptor dose) are properly integrated with the overall system model.
22   The reviewer should ensure that the performance assessment models are developed with
23   proper integration of models for each of the exposure scenarios (e.g., onsite, offsite, and
24   inadvertent intruder scenarios). If a probabilistic model is used, the reviewer should ensure that
25   the number of model realizations used to estimate expected dose to a receptor is sufficient to
26   achieve a stable mean dose estimate (i.e., results do not change significantly if more model
27   realizations are simulated). Additionally, the reviewer should ensure that the numerical
28   accuracy of model calculations has been verified as part of the code development process.
29

30   **4.4.2   Review Procedures**

31
32   •   The reviewer should determine the adequacy and completeness of the quality assurance
33   documentation of the code/model (see Section 8). This review should include
34   documentation pertaining to (1) software requirements and intended use, (2) software
35   design and development, (3) software design verification, (4) software installation
36   and testing, (5) configuration control, (6) software problems and resolution, and
37   (7) software validation.
38
39   •   The reviewer should ensure that the computational model is compatible with the disposal
40   site conceptual model, including the pathways and the exposure scenario. The
41   source-term assumptions of the selected code should also be compatible with the
42   site-specific source term. For example, if the code selected for source-term analysis
43   in the performance assessment does not estimate diffusive releases, the reviewer
44   should ensure that diffusive releases are not important to the performance of the
45   disposal facility.
46
47   •   The reviewer should evaluate information on the limitations of the selected codes to
48   ensure that modeling results and modeling approaches are not being arbitrarily
49   constrained by the limitations of the codes selected for the analysis. For example, the
50   selected code may not be able to model the effects of preferential pathways, model

1   certain exposure pathways, or represent changes in model parameters as a function of
2   time (e.g., for barrier degradation).
3

4 •   The reviewer should evaluate the code documentation to verify that the exposure
5   scenarios of the performance assessment code are compatible with the intended
6   scenarios for the site.
7

8 •   The reviewer should verify that unit conversion errors have not occurred as a result of
9   the passage of information between models in the analysis.
10

11 •   The reviewer should ensure that contaminant fluxes have not been arbitrarily dispersed
12   (e.g., numerically) or diluted artificially in the performance assessment modeling
13   (e.g., artificial dilution of fluxes into larger than expected depths of the saturated zone
14   because of the size of finite elements in the groundwater model). A minor amount of
15   numerical dispersion may be expected.
16

17 •   If a probabilistic analysis is performed, the reviewer should ensure that an adequate
18   number of stochastic realizations have been executed to produce stable output.
19

20 •   If a deterministic analysis is performed, the reviewer should determine whether the
21   overall performance assessment is sufficiently conservative to account for uncertainty in
22   the parameters and models. For example, the reviewer can make a qualitative,
23   semiquantitative, or quantitative comparison of a list of conservative assumptions or
24   approaches in the analysis with a list of key parameters that were represented as
25   deterministic in the analysis but would be expected to be uncertain.
26

27 •   The reviewer should ensure that the performance assessment model results are not
28   sensitive to time stepping or spatial discretization of the model domain.
29

30 •   The reviewer should ensure the performance assessment codes properly account for
31   radionuclide decay and ingrowth. The codes should provide a mechanism to track the
32   amounts of radioactivity throughout the disposal system and site over the performance
33   period, including which environmental media the radionuclides are present in or
34   associated with.
35

36 •   The reviewer should ensure that the treatment of the transport properties assumed for
37   daughter radionuclides created during a transport leg do not underestimate dose. For
38   example, many dose modeling codes do not assign new properties to daughter
39   radionuclides that are created during transport and can overestimate transport times for
40   daughters that are more mobile than the parent (e.g., Np-237 ingrowth from Am-241).
41

42 **4.5   Uncertainty/Sensitivity Analysis for Overall**
43   **Performance Assessment**
44

45 **4.5.1   Areas of Review**
46

47 General approaches to handling data and model uncertainty are described in Section 4.4.1.1.
48 As discussed previously, probabilistic approaches to performance assessment are preferred in
49 most cases because they readily permit the propagation and assessment of the impact of

1 uncertainty on the model results. However, use of a deterministic model to demonstrate
2 compliance with performance objectives is acceptable. In general, if deterministic modeling is
3 used, it should be reasonably conservative such that a subject matter expert, with minimal
4 interaction with those who performed the assessment, could conclude that the analysis
5 was conservative.
6
7 For modeled processes and input parameters that are highly uncertain and cannot clearly be
8 established as conservative, sensitivity analyses are necessary to establish the relative
9 importance of these processes and parameters to the performance assessment dose
10 calculations. A summary of different methods for sensitivity and uncertainty analyses can be
11 found in NUREG–1757 (NRC, 2003, Vol. 2, Appendix I, Section I.7) and in NUREG–1573 (NRC,
12 2000, Section 3.3.2).
13
14 As discussed previously, different approaches to performance assessment calculations
15 (e.g., deterministic, probabilistic) have their advantages and disadvantages with regard to
16 uncertainty and sensitivity analysis. While deterministic analysis can be a suitable approach for
17 performance assessment, it can also present a challenge for a dynamic system that responds
18 nonlinearly to the independent variables. When there are numerous inputs (e.g., data or
19 models) that are uncertain, evaluating the impacts of the uncertainties on the decision can be
20 difficult. Typical one-off type of sensitivity analysis where a single parameter is increased or
21 decreased will only identify local sensitivity within the parameter space such that it may not
22 clearly identify the risk implications. A deterministic approach can be useful to bound
23 uncertainty when the analysis can be demonstrated to be conservative. A probabilistic
24 approach can have distinct advantages when there are a number of uncertainties that may
25 significantly influence the results of a performance assessment. For example, the uncertainty
26 introduced by the changing effectiveness of a chemical barrier over time may be represented by
27 selecting appropriate ranges in the transport parameters for the barrier. If a probabilistic
28 approach is used for the performance assessment, the reviewer will need to determine whether
29 there is significant "risk dilution" affecting the calculation results. Risk dilution results when
30 overly broad parameter distributions are selected, primarily for processes that affect the timing
31 of impacts that are expected to occur in the performance period. For example, selection of an
32 overly broad range for the distribution coefficient for Tc-99 in the saturated zone may result in
33 the estimated time of arrival of the contaminant at a receptor location being artificially extended
34 over the period of performance, thereby "diluting" the risk at any one point in time.
35
36 If DOE has performed a deterministic performance assessment, then the reviewer should
37 examine the sensitivity analyses DOE provided. Review the basis for selecting the parameters
38 and combinations of parameters used in the sensitivity analysis. The ranges in the parameters
39 selected should be consistent with the variability and uncertainty in the parameters, and the
40 selected ranges should provide the reviewer confidence that the effects of the uncertainty on
41 performance are bounded. The reviewer should examine the technical basis used to support
42 the variability and uncertainty. Appropriate combinations of parameters should be used to
43 capture the interdependence of key parameters. The reviewer should consider combinations of
44 parameter values that are likely to occur as a result of common causes.
45
46 Key parameters for further evaluation in sensitivity analysis may be defined with a variety of
47 different approaches, and the appropriateness of the approaches is problem dependent. It is
48 anticipated that the sensitivity analysis and the performance assessment overall may be an
49 iterative process. The initial approach evaluated may not be the final approach selected.
50 Regardless of the process, the reviewer should keep in mind that the purpose of the sensitivity

1 analysis is to evaluate uncertainty and variability in the assessment. One of the simplest
2 methods utilizes a top down approach where risk reduction of each component (e.g., infiltration,
3 unsaturated zone, wasteform, engineered barriers, saturated zone, biosphere characteristics) of
4 the performance assessment model is identified by starting with a hazard and calculating how
5 each component reduces the risk from the hazard. Subsequently, within the most important
6 components, a quantitative or qualitative evaluation of the parameters is performed to identify
7 those that are most likely to influence the output from the component. Complications arise
8 because an individual component's importance in the system can be relative to the performance
9 of other components. Sections 4.6.1.1 and 4.6.1.2 provide more detail on the process of
10 evaluating models. Developing an understanding of the importance of parameters and models
11 in a performance assessment is a time-consuming process that is best accomplished by
12 exploring a variety of approaches.
13
14 Because conceptual models are sometimes developed based on limited data, in some cases
15 more than one possible interpretation of the site can be justified based on the existing data.
16 This uncertainty should be addressed by developing multiple alternative conceptual models and
17 implementing the conceptual model that most conservatively estimates the dose or best
18 represents the available data. Alternatively, DOE could provide analyses for multiple
19 conceptual models to develop a range of dose estimates. Consideration of unrealistic and
20 highly speculative conceptual models should be avoided. Consistent with the overall dose
21 modeling framework of starting with simple analyses and progressing to more complex
22 modeling, as warranted, the analyst may want to begin with a simple, conservative analysis that
23 incorporates the key site features and processes and progress to more complexity only as site
24 data merit. A simple representation of the site, in itself, does not mean that the analysis is
25 conservative. The reviewer should evaluate whether DOE has presented information
26 demonstrating that its simplification is justified, based on what is known about the site and the
27 likelihood that alternative representations of the site would not lead to higher calculated doses.
28

29 **4.5.2   Review Procedures**
30
31 •   For deterministic analysis, evaluate the adequacy of the process for determining key
32   model parameters.
33
34 •   If a deterministic model framework is adopted, ensure that key model parameters are
35   either conservative, well justified, or that sensitivity analyses have been provided to
36.   demonstrate the overall risk significance of the parameters.
37
38 •   If conservatism is used as a basis to select deterministic parameter values for the
39   sensitivity analysis, evaluate whether conservatism has been defined locally or globally.
40   For example, stating that the selected saturation value is conservative because it has
41   been set to 1.0 (the highest value expected) and that a saturation of 1.0 maximizes
42   contaminant transport because it maximizes relative permeability is a local argument if a
43   cementitious wasteform is involved. Carbonation of cements is maximized at
44   intermediate values of saturation (due to transport of carbon dioxide in the gas and
45   liquid phases). Carbonation of a cementitious wasteform may eventually disrupt
46   the engineered barrier, resulting in higher release rates compared to assumed
47   saturated conditions.
48
49 •   If the performance assessment is complex, verify that the technical basis for the
50   conservatism of the sensitivity analysis is supported in at least a semiquantitative

fashion, such as a diagram showing the parameters being evaluated and which submodels of the performance assessment that the parameters affect and in what way. A complex performance assessment would have characteristics such as but not limited to (1) integrated submodels that respond nonlinearly to changes in parameters, (2) many uncertain parameters, (3) high temporal and spatial variability of processes and properties, and (4) many integrated submodels.

- Evaluate the ranges of parameters used in sensitivity analysis, and verify that they reasonably cover the expected ranges.

- Ensure that appropriate combinations of parameters have been used in sensitivity analysis to capture the interdependence of key parameters. For example, consider (1) combinations of parameters affecting radionuclide release that may be physically coupled, such as increased infiltration that may result in more rapid engineered barrier and wasteform deterioration and increased release rates from the wasteform; (2) combinations of parameters used to represent processes that would be expected to occur together, such as cracking of a wasteform due to seismic activity that may affect both physical and chemical durability of the wasteform; and (3) combinations of parameters affecting transport such as decreased K_d values for chemically similar radionuclides due to an external factor (e.g., leaching of alkalinity from a disposal facility).

- If a deterministic analysis with a sensitivity analysis has been performed and if resources allow, perform an independent probabilistic assessment to evaluate the DOE analysis.

- If DOE has adopted a probabilistic framework, evaluate whether it is sufficient to support understanding of the overall risk significance of uncertain model input parameters.

- If DOE has adopted probabilistic analyses, the reviewer should ensure that unrealistic ranges in parameter distributions that could result in "risk dilution" are not used. This is especially important when the parameter distribution is assigned based on generic or literature information and when the parameter influences the timing of the occurrence of the peak dose (e.g., K_d distributions).

- As appropriate, ensure that DOE has considered alternative conceptual models and that demonstrations of compliance with performance objectives are based on well-supported conceptual models or the most conservative model that is consistent with site characteristics.

4.6   Evaluation of Model Results

4.6.1   Areas of Review

4.6.1.1   Defining Barrier Contributions

The prior sections of this guidance document address the review of detailed descriptions of engineered and natural barriers and their implementation in the performance assessment model. The reviewer should ensure that the DOE discussion of performance assessment results includes a quantitative and qualitative analysis and description of how engineered and

1 natural barriers in the performance assessment model limit or prevent doses. The reviewers
2 should use information defining barrier contributions to risk-inform their review.
3
4 The term "barrier" as used in this context, applies to engineered or natural components of the
5 system that may reduce or mitigate risks. Engineered components may be those components
6 specifically designed for the waste disposal facility (e.g., an infiltration cap), or in the case of
7 closure of a waste storage facility, those components of the waste storage facility not
8 specifically designed for long-term performance but that may affect the long-term performance
9 of the system (e.g., a vault holding an underground tank). Results may be provided for intact
10 versus failed (or realistically degraded) barrier performance for individual barriers (and
11 collections of barriers) to gain insights into the importance of barriers on the performance
12 assessment results. In addition, results may be presented with only the natural system or only
13 the engineered system present in the analysis to show the relative contribution of the
14 engineered and natural systems to overall performance of the waste disposal facility. Without
15 an analysis of barrier contributions, the influences of barriers on performance assessment
16 results may not be transparently described or self-evident. In this regard, staff should verify that
17 DOE adequately describes the contributions of key engineered and natural barriers to the
18 performance assessment results. The reviewer should ensure that information about the barrier
19 functions, including the magnitude of the impact of the barrier function on estimated doses, is
20 reasonable and consistent with the descriptions of the engineered barriers (see Section 4.3.2)
21 and characteristics of the disposal site.
22
23 **4.6.1.2   Evaluating Intermediate Model Results**
24
25 The main purpose of the performance assessment model is to estimate potential long-term
26 radiological dose to receptors. Typically, a large amount of intermediate outputs (such as
27 infiltration through a cap or the fractional release rate from a wasteform) is produced during
28 execution of a performance assessment model. In addition to evaluating the contribution of
29 engineered barriers to reduce or mitigate radiological dose, the reviewer should evaluate
30 intermediate outputs of performance assessment models to understand the interactions of
31 barriers and any possible masking effects of barriers. For example, the infiltration rate through
32 an engineered cap may not be risk significant if the hydraulic properties of the wasteform limit
33 flow to lower values. However, in this case, the engineered cap may be providing a redundant
34 performance function, and it may also be contributing to mitigating degradation mechanisms of
35 the wasteform associated with water flow (e.g., leaching). As appropriate, the reviewer should
36 examine the intermediate results of the performance assessment calculations to understand the
37 integration of the performance assessment models.
38
39 **4.6.1.3   Final Dose Calculations**
40
41 Numerous NRC guidance documents recommend approaches and specific dose conversion
42 factors used in performance assessments. These include NUREG–1573 (NRC, 2000), which
43 provides guidance on the use of pathway dose conversion factors for calculating doses via
44 potential exposure pathways, and NUREG–1757 (NRC, 2003, Vol. 2, Appendix I), which
45 provides guidance on the use of dose conversion factors such as those developed by the
46 U.S. Environmental Protection Agency (EPA) and published in Federal Guidance Reports 11
47 and 12 (EPA, 1988, 1993).
48

1   The reviewer should assess the dose conversion factors for inhalation and ingestion to
2   ensure that the factors used are those developed by EPA, published in Federal Guidance
3   Report No. 11 (EPA, 1988). Similarly, the reviewer should ensure that the EPA external dose
4   factors published in Federal Guidance Report No. 12 (EPA, 1993) were used. These dose
5   factors were selected to ensure consistency of the dosimetry models used in deriving these
6   factors with NRC regulations in 10 CFR Part 20. The reviewers should evaluate the dose
7   conversion factors used to analyze external doses to ensure the factors are appropriate for the
8   thickness, shielding, and extent of contamination.
9

10   This guidance document advocates the use of dosimetry consistent with 10 CFR Part 20. This
11   ensures that compliance calculations between 10 CFR 61.41 and 10 CFR 61.43 remain
12   consistent, as discussed in NUREG–1573. The proposed rule for 10 CFR Part 63, "Disposal of
13   High-Level Radioactive Wastes in a Proposed Geological Repository at Yucca Mountain,
14   Nevada; Proposed Rule," states
15

16   "As a matter of policy, NRC considers 0.25 mSv [25 mrem] TEDE as the
17   appropriate dose limit within the range of potential doses represented by the
18   current 10 CFR 72.104 limit of 0.25 mSv [25 mrem] (whole body), 0.75 mSv
19   [75 mrem] (thyroid dose), and 0.25 mSv [25 mrem] (to any other critical organ)
20   (NRC 1999)."
21

22   As 10 CFR 61.41 has the same standard as 10 CFR 72.104, this policy is applicable, and
23   therefore, incidental waste determinations may use total effective dose equivalent (TEDE)
24   without specific consideration of individual organ doses. Intruder calculations should be based
25   on 5 mSv [500 mrem] TEDE limit, without specific consideration of individual organ doses, to
26   ensure consistency between 10 CFR 61.41 and 10 CFR 61.43. Because of the tissue weighting
27   factors and the magnitude of the TEDE limit, specific organ dose limits are not necessary for
28   protection from deterministic effects.
29

30   DOE may use the latest dose conversion factors (e.g., ICRP–72). While not completely
31   identical in detail, NRC would generally consider 0.25 mSv [25 mrem] effective dose to be the
32   same as 0.25 mSv [25 mrem] TEDE for these problems. However, the use of dose conversion
33   factors for individual radionuclides or by pathways (I.e., DOE should use the alternate dosimetry
34   for all parts of the analysis [10 CFR 61.41, 61.43, and intruder analyses] and for all
35   radionuclides). The exception to this statement is for direct exposure pathways, which were not
36   updated in ICRP–72. Therefore, older dose conversion factors for direct exposure pathways
37   may be combined with newer dose conversion factors for inhalation and ingestion pathways.
38   Scenarios and critical group assumptions should account for age-based considerations.
39

40   **4.6.1.4   Comparison to Performance Objectives**
41

42   The postclosure performance assessment is used to demonstrate compliance with
43   10 CFR 61.41, "Protection of the general population from releases of radioactivity,"
44   which states
45

46   "Concentrations of radioactive material which may be released to the general
47   environment in groundwater, surface water, air, soil, plants, or animals must not
48   result in an annual dose exceeding an equivalent of 25 millirems to the whole
49   body, 75 millirems to the thyroid, and 25 millirems to any other organ of any
50   member of the public. Reasonable effort should be made to maintain releases

1   of radioactivity in effluents to the general environment as low as is
2   reasonably achievable."

4   The 0.25 mSv/yr [25 mrem/yr] limit applies for the postclosure period of a disposal facility. NRC
5   expects DOE to express this limit in TEDE (NRC, 2005, 1999, Footnote 1). For probabilistic
6   performance assessment models, it is acceptable to use the peak of the mean dose history to
7   demonstrate compliance with this performance objective.

9   The reviewer should evaluate that the appropriate dose limits are applied. To evaluate whether
10   the performance objectives in 10 CFR 61.41 can be met, probability can be considered in the
11   consequence analysis, and the annual dose limit to the member of the general public is
12   0.25 mSv [25 mrem] (NRC, 2005, 1999, Footnote 1). For example, disposal facility
13   performance may be severely affected by a low-frequency, large-magnitude seismic event. In
14   this case, it would be appropriate to factor in the event probability when estimating the risk to
15   the public. In estimating dose to an inadvertent intruder (see Section 5), however, the
16   probability of the intrusion in the stylized analysis is assumed to be 1, and a higher annual dose
17   limit of 5 mSv [500 mrem] is assumed for an inadvertent intruder (NRC, 1981).

19   The limit provided in 10 CFR 61.41 {0.25 mSv/yr [25 mrem/yr]} applies to the cumulative
20   impacts from the LLW disposal units that could contribute to the receptor dose (e.g., tanks in a
21   tank farm subject to past, present, and future waste determinations). Because the amount of
22   waste associated with future determinations will not be known in the present, the reviewer will
23   need to consider estimated doses in previous waste determinations for different sources
24   (i.e., those not part of the current waste determination) that could contribute to receptor doses.
25   Eventually, demonstration of compliance with 10 CFR 61.41 will be based on the maximum
26   dose to a receptor from all disposal units subject to waste determinations. All disposal units
27   may not contribute to receptor doses at a particular location (e.g., groundwater flow paths may
28   be in different directions due to the presence of a groundwater divide). The reviewer should
29   note in the TER the cumulative impact from past and current waste determinations.

31   Demonstration of stability of the disposal site after closure requires a separate technical
32   evaluation, and review procedures for this performance objective are described in Section 7 of
33   this guidance document. Reviewers of the postclosure performance assessment models should
34   evaluate whether the site stability performance objective is met because long-term site stability
35   is typically an important assumption in the performance assessment conceptual model.

4.6.2   Review Procedures

39   The following review procedures focus on ensuring that performance assessment model results
40   are sufficient for comparison to performance objectives. The reviewer should perform the
41   following review procedures:

43   •   Examine the descriptions of the methodology used for conversion of internal and
44   external exposure to dose.

46   •   Ensure the dose conversion factors or correction factors DOE used in the analysis of
47   external doses are appropriate for the thickness, shielding, and extent of contamination.

49   •   Ensure that the dose conversion factors selected are consistent with the dosimetry
50   model (e.g., FGR 11 and 12 values are used with ICRP 26 and 30).

- If ICRP–72 values are used, ensure the alternate dosimetry is used for all parts of the analysis (10 CFR 61.41, 61.43 and intruder analyses) and for all radionuclides. An exception would be to use FGR 12 values for direct exposure pathways while using ICRP–72 values for inhalation and ingestion, because the direct exposure values were not updated in ICRP–72.

- Ensure that solubility classes were conservatively selected or technical basis has been provided for selecting less conservative values.

- Review the DOE description of the comparison of performance assessment model output to the applicable performance objectives in 10 CFR Part 61, Subpart C. Ensure that the description includes an appropriate performance measure such as peak of the mean dose for probabilistic assessments or peak dose for deterministic assessments.

- Ensure that descriptions of performance assessment results are sufficient to permit comparison of model results to the performance objective of 10 CFR 61.41.

- Ensure that DOE is using the appropriate dose limit from 10 CFR 61.41 for the public during the period after active institutional controls.

- Ensure that probabilities have been appropriately applied to the determination of whether the performance objective in 10 CFR 61.41 can be met, if necessary.

- Evaluate compliance with 10 CFR 61.41 by ensuring that the maximum cumulative dose to a public receptor from disposal units evaluated in past waste determinations and the present waste determination does not exceed 0.25 mSv/yr [25 mrem/yr]. The reviewer should assess whether the location of the receptor that is expected to receive the maximum cumulative dose was appropriately determined or bound (e.g., impacts from each disposal unit were assumed to be along the same groundwater flow path).

- Evaluate the sensitivity/uncertainty analyses provided with the performance assessment to determine whether the analyses are sufficient to permit evaluation of model sensitivity to uncertain processes and input parameters.

- Verify that modeled processes and input parameters are treated in a manner that is either demonstrably conservative, sufficiently constrained by site data, or otherwise demonstrated to contribute to an insignificant uncertainty effect on modeled dose estimates.

- Ensure that DOE has addressed uncertainty in the timing of the peak dose if a deterministic analysis is used. The reviewer should evaluate whether significant peak doses occur after the period of performance and whether uncertainties in any of the transport or release models or parameters would be sufficient to move the peak doses into the performance period.

- Ensure that DOE has provided an adequate technical basis to explain differences between current and past modeling of the disposal facility, as appropriate.

- Verify that DOE has provided a complete description of all the key barriers included in the performance assessment model. This review should be integrated with the detailed technical reviews conducted in Sections 2–8.

- Verify that the DOE description and analysis of performance assessment results provides a complete (qualitative and quantitative) analysis and description of barrier contributions to performance assessment results. The discussion of results should include a transparent description of how the modeled disposal site system is functioning, with an emphasis on barriers to release and transport that affect the magnitude or timing of estimated doses. A brief example of the type of information that may be included in such a barrier performance analysis description is provided in the following paragraph:

The engineered cap reduces infiltration to the waste from 25 cm/yr [10 in/yr] to 1 cm/yr [0.4 in/yr] for 1,000 years. Without the engineered cap present, the dose only increased a factor of 2 from the nominal case because of the low hydraulic properties of the wasteform. In the case of the cap failing and the wasteform degrading hydraulically, the dose is increased by a factor of 20 from the nominal case. The chemical properties of the wasteform provide a significant barrier to radionuclide release, even if the system does not perform as intended from a hydrologic standpoint. Without the chemical properties of the wasteform, the dose would increase an additional factor of 100, primarily as a result of the release of relatively strongly sorbing Np-237 and Pu-239.

- Following review of the DOE description of performance assessment results, ensure an adequate technical basis is provided for barrier capabilities that contribute significantly to performance assessment results (i.e., those barriers that significantly reduce estimated doses or significantly delay the estimated doses). The characteristics of the barriers should be described in detail to provide confidence in the performance capabilities of the barriers.

- Evaluate whether the DOE performance assessment has appropriately accounted for applicable barrier interactions (e.g., change or variation in chemistry that degrades an engineered barrier could also influence sorption or source term release estimates). Analysis of the performance of individual barriers one at a time may be incomplete, because this type of analysis may miss plausible and potentially important interactions of features and processes that can amplify impacts on system performance.

- Evaluate whether the modeling of barriers in the performance assessment appropriately accounts for and propagates applicable uncertainties and variabilities or (in particular for deterministic analyses) that barriers are represented conservatively.

- Verify that the description of each barrier is consistent with the detailed description and supporting information for barrier features and capabilities.

- Evaluate intermediate results from the DOE performance assessment to understand the interactions of barriers and any possible masking effects of barriers. Verify that intermediate results (e.g., fluxes, travel times) are physically reasonable.

4.7 ALARA Analysis

This part of the review focuses on assessing compliance with the requirement in 10 CFR 61.41 that releases of radioactivity in effluents from the disposal facility to the general environment be maintained as low as is reasonably achievable (ALARA). The review of the postclosure performance assessment to determine compliance with the dose requirements of 10 CFR 61.41 is described in Sections 4.1-4.6. Review of the performance objective for maintaining radiation exposures to individuals during operations ALARA, as required in 10 CFR 61.43, is discussed in Section 6 of this guidance document.

4.7.1 Areas of Review

In general, the conclusion that proposed waste management activities will result in the removal of highly radioactive radionuclides to the maximum extent practical (see Section 3) supports the conclusion that releases of radioactivity in effluents from the disposal site will be maintained ALARA. Thus, a reviewer should begin the evaluation of compliance with the ALARA requirement of 10 CFR 61.41 by completing the review described in Section 3 of this guidance document. In addition, because steps taken to stabilize waste also are expected to limit radionuclide release from the disposal facility, stabilization activities also are relevant to the assessment of compliance with the requirement to maintain effluents ALARA. Therefore, a reviewer should evaluate DOE's description of actions taken to stabilize the waste to minimize the release of radionuclides from the disposal facility (e.g., efforts to optimize solidification of liquid wastes or efforts to optimize mixing or encapsulation of residual tank waste with grout). The review should be focused on the dominant pathways of radionuclide release from the disposal facility and the factors causing the most uncertainty in release rates, as determined in DOE's performance assessment and independent analyses.

4.7.2 Review Procedures

After completing the review described in Section 3 of this guidance document, the reviewer should identify the dominant pathways of radionuclide release from the disposal site based on the results of DOE's performance assessment and any independent analysis used in the review described in Sections 4.1-4.6 of this guidance document. When identifying dominant release pathways, the reviewer should also identify key uncertainties that may affect which factors dominate radionuclide release (e.g., placement of waste in areas thought to be especially prone to the development of preferential pathways for infiltrating water may cause significant uncertainty in releases to groundwater). In general, the reviewer should determine what steps have been taken to limit the release of radionuclides through the dominant release pathways and to limit the potential for high release rates due to alternate release mechanisms. Specifically, the reviewer should perform the following activities:

- Confirm that DOE has provided sufficient support for the performance assessment that the dose requirements of 10 CFR 61.41 can be met (see Sections 4.1-4.6).

- Determine whether waste is placed in areas that are susceptible to the formation of preferential pathways for infiltrating water (e.g., along joints between dissimilar materials) and whether appropriate steps have been taken to move waste from those areas, if practical.

1 • Determine whether appropriate efforts have been made to optimize the solidification of
2 any relevant liquid waste (in general, wastes containing more than 1 percent residual
3 liquid by volume are not suitable for near-surface disposal [10 CFR 61.56]).
4
5 • If stabilizing material is added to a waste, determine whether appropriate efforts have
6 been made to optimize mixing or encapsulation of the waste with the stabilizing material
7 (e.g., evaluate the selection of methods used to optimize mixing of residual waste and
8 grout in tanks).
9
10 To the extent possible, the determination of whether efforts to optimize waste stabilization are
11 appropriate should be evaluated based on quantitative measures, using procedures similar to
12 those discussed in Section 3. For example, if releases from wastes located along the edges of
13 a tank contribute significantly to releases of radionuclides into groundwater or contribute
14 significant uncertainty to the expected releases from the site, the reviewer should evaluate
15 DOE's selection of options available to move or stabilize the wastes present along the edge of
16 the tanks. Procedures similar to those used to evaluate DOE's selection of technologies
17 available to remove the waste from the tank, as described in Section 3, should be considered.
18 Similarly, to determine whether actions that could stabilize waste are practical, it may be
19 necessary to compare the costs and benefits of additional stabilization, as described in
20 Section 3.4.
21

22 ## 4.8 References
23

24 Bradbury, M. and F.A. Sarott. "Sorption Databases for the Cementitious Near-Field of a L/ILW
25 Repository for Performance Assessment." PSI Bericht 95-06, Wurenlingen and Villigen,
26 Switzerland: Paul Scherrer Institut. 1995.
27
28 U.S. Environmental Protection Agency (EPA). "Federal Guidance Report No. 11: Limiting
29 Values of Radionuclide Intake and Air Concentration and Dose Conversion Factors for
30 Inhalation, Submersion, and Ingestion." EPA–520/1–88–020. September 1988.
31
32 ———. "Federal Guidance Report No. 12: External Exposure to Radionuclides in Air, Water
33 and Soil." EPA–402–R–93–081. September 1993.
34
35 ———. "Understanding Variation in Partition Coefficient, K_d, Values." EPA-402-R-99-004A.
36 August 1999.
37
38 U.S. Nuclear Regulatory Commission (NRC). "Draft Environmental Impact Statement on
39 10 CFR Part 61 Licensing Requirements for Land Disposal of Radioactive Waste."
40 NUREG–0782. September 1981.
41
42 ———. "Final Environmental Impact Statement on 10 CFR Part 61 Licensing Requirements for
43 Land Disposal of Radioactive Waste." NUREG–0945. 1982.
44
45 ———. "Generic Technical Position on Peer-Review for High-Level Nuclear Waste
46 Repositories." NUREG–1297. February 1988a.
47
48 ———. "Generic Technical Position on Qualification of Existing Data for High-Level Nuclear
49 Waste Repositories." NUREG–1298. February 1988b.

1   ——. "Branch Technical Position on the Use of Expert Elicitation in the High-Level
2   Radioactive Waste Program." NUREG–1563. November 1996.
3
4   ——. "Disposal of High-Level Radioactive Wastes in a Proposed Geological Repository at
5   Yucca Mountain, Nevada." Proposed Rule. *Federal Register,* 64 FR 8640. February 1999.
6
7   ——. "A Performance Assessment Method For Low-level Waste Disposal Facilities:
8   Recommendations of NRC's Performance Assessment Working Group." NUREG–1573.
9   October 2000.
10
11   ——. "Consolidated NMSS Decommissioning Guidance." Vols. 1–3. NUREG–1757.
12   September 2003.
13
14   ——. "Technical Evaluation Report for the U.S. Department of Energy, Savannah River Site
15   Draft Section 3116 Waste Determination for Salt Waste Disposal." Letter from L. Camper to
16   C. Anderson, DOE. December 2005.
17

5 INADVERTENT INTRUSION

The performance objective for protection of the inadvertent intruder is provided in 10 CFR 61.42. Specifically, the staff review should confirm whether the design, operation, and closure of the land disposal facility will ensure protection of an individual inadvertently intruding into the disposal site and occupying the site or contacting the waste at any time after active institutional controls for the disposal site are removed. The performance objective in 10 CFR 61.42 does not provide a numerical dose criteria for protection from the inadvertent intruder. In previous waste determination reviews (NRC, 2000a, 2003a, 2005), the U.S. Nuclear Regulatory Commission (NRC) has applied the whole body-dose equivalent limit of 5 mSv/yr [500 mrem/yr] described in Section 4.5 of Volume 2 from the Draft Environmental Impact Statement for 10 CFR Part 61 (NRC, 1981) to assess intruder scenarios. When evaluating protection of the inadvertent intruder, NRC assumed that active institutional controls would be maintained for 100 years following permanent closure [10 CFR 61.59(b)].

Additional guidance on intruder analysis can be found in NUREG–0782 (NRC, 1981, Vol. 2, Chapter 4), NUREG–1200 (NRC, 1988), and NUREG/CR–4370 (Oztunali and Roles, 1986). Performance assessment modeling approaches are discussed in Section 4 of this guidance document, and additional guidance for developing performance assessment and dose assessment models can be found in NUREG–1573 (NRC, 2000b) and NUREG–1757 (NRC, 2003b). Specific guidance on the period for active institutional controls is provided in Section 4.1.1.1, and the influence of regional practices on scenario identification is discussed in Sections 4.1.1.3 and 4.1.1.4. Protection of individuals during operations is discussed in Section 6, and site stability is discussed in Section 7 of this guidance document.

5.1   Areas of Review

In evaluating the intruder protection system proposed by the U.S. Department of Energy (DOE), reviewers should consider the operations, procedures, materials, barriers, and structures designated to provide this protection. Active protection systems may be effective during a 100-year period of institutional control, but given the long time periods considered in performance assessment used to predict potential doses following permanent closure of the facility, reviewers may need to consider the performance of engineered and natural systems as passive barriers to intrusion. A passive barrier is a barrier that may perform its intended functions without active monitoring and maintenance. The reviewer should take a risk-informed, performance-based approach to reviewing any site-specific intruder analysis and inadvertent intruder protection systems DOE proposed, and the reviewer should evaluate the parameters used in the intruder analysis in the context of regional practices. This approach provides DOE with flexibility in analyzing intruder scenarios to demonstrate compliance with the performance objectives of 10 CFR 61.42.

5.1.1   Assessment of Inadvertent Intrusion

The reviewer should assess DOE-provided information regarding estimates of the potential dose to an inadvertent intruder. Because many intruder scenarios may be simpler and involve less integration of submodels compared to the public dose scenarios evaluated for compliance with 10 CFR 61.41, a more simplified analysis could be performed. Intruder scenarios commonly involve disruption of the waste disposal facility and exhumation of waste, with resultant calculations of dose. The reviewer will need to consider the complexity of the intruder

1 scenarios and determine whether the analysis was sufficiently detailed to account for important
2 features, events, and processes. The review of the nominal case performance assessment is
3 described in Section 4 of this guidance document. The reviewer should consider the guidance
4 found in Section 4, as needed, for review of inadvertent intruder scenarios. In most cases, the
5 reviewer will only need to consider the guidance found in this section for inadvertent intruder
6 analysis. The reviewer should evaluate the following information regarding inadvertent intrusion:
7
8 • Technical bases and associated analyses used to define the intruder scenarios. This
9 should include aspects such as the behavior of the intruder, timing of the intrusion, and
10 the exposure pathways simulated in the intruder analysis.
11
12 • Assumptions and parameters used in developing the intrusion assessment,
13 characteristics of the intrusion event, how uncertainties are considered in the analysis,
14 and resultant conditional doses for each intruder scenario. The assumptions in the
15 intruder analysis should be consistent with past, current, and projected regional
16 practices and the performance assessment (see Section 4).
17

5.1.2 Intruder Protection Systems

18
19
20 As discussed in Section 4.1.1.1, active institutional controls are limited to 100 years or less after
21 closure, but DOE may propose passive systems to protect from inadvertent intrusion. The
22 reviewer should examine any DOE-proposed intruder protection system. This could include
23 engineered barriers (caps, rock layers) or waste system placement and burial (e.g., depth to
24 waste) to reduce the potential for inadvertent intrusion. As specified in 10 CFR Part 61, intruder
25 barriers should prevent or limit intrusion for 500 years for Class C waste. After 500 years, the
26 activity of Class C waste would decay such that an inadvertent intruder would be protected if
27 they contacted the waste. The service life for an intruder barrier for greater than Class C waste
28 may need to be considerably longer than for Class C waste, and technical justification for
29 long-term performance may be considerably more challenging. As discussed in Section 3.2.2 of
30 NUREG–1573, service lives for engineered barriers, on the order of a few hundred years, are
31 considered credible if justified by adequate technical analyses and data (NRC, 2000b). The
32 reviewer should review how the features of the engineered barriers and projected service
33 lifetimes are incorporated in the intruder analysis. Depth to waste is an important consideration
34 in defining intruder scenarios, because some scenarios may not be credible if the waste is
35 generally inaccessible. The reviewer should review the projected depth to waste for the waste
36 disposal system being evaluated and how the depth may change during the performance period
37 (e.g., because of erosion).
38

5.1.3 Types of Scenarios Considered in the Intruder Analysis

39
40
41 Future human behavior cannot be accurately predicted over hundreds to thousands of years.
42 To address this uncertainty, hypothetical intruder scenarios are designed to bound the exposure
43 to the intruder while avoiding speculation about future human activities. The regulations in
44 10 CFR 61.42 do not specify that a particular scenario be used to demonstrate compliance. In
45 developing intruder scenarios, it is anticipated that DOE will assume humans will continue
46 normal land use activities that are consistent with recent past (e.g., a few decades) and current
47 regional practices after active institutional controls are no longer enforced. DOE is not expected
48 to provide probability estimates for the individual scenarios, but conditional doses should be
49 estimated for each scenario considered in the inadvertent intruder analysis. To the extent that

DOE has provided an analysis of various intruder scenarios, the reviewer should evaluate the site-specific scenarios using the guidance in Section 5.2. This may include one or more of the following scenarios.

5.1.3.1 Intruder-Resident Scenario

This scenario assumes that after the end of active institutional controls, an intruder (i.e., the resident intruder) inadvertently constructs a house at and lives on the waste disposal area. The reviewer should assess the location of the resident intruder relative to the waste disposal system, the behavior attributed to the intruder, and the timing of the intrusion. The reviewer should examine dose pathways assumed for this scenario (e.g., direct exposure, ingestion [drinking water, vegetables, soil], inhalation) and the parameters and calculations used to estimate intruder doses. The intruder-resident is typically assumed to have a garden that is used to raise produce (for chronic exposures). The resident intruder may also be exposed to contaminated drill cuttings that resulted from installation of a well for domestic purposes (see Section 5.1.3.4).

5.1.3.2 Intruder-Agriculture Scenario

This scenario assumes that after the end of active institutional controls, a farmer lives on and consumes food crops grown and animals raised on the disposal area. The reviewer should assess the location of the intruder relative to the waste disposal system, the behavior attributed to the intruder, and the timing of the intrusion. The reviewer should examine dose pathways assumed for this scenario (e.g., direct exposure, ingestion [drinking water, plant products, animal products, soil], inhalation) and the parameters and calculations used to estimate intruder doses. The intruder-agriculture is typically assumed to have a garden that is used to raise produce (for chronic exposures). The reviewer should examine the extent to which the intruder analysis accounts for ground-disturbing activities by the farmer (e.g., plowing, spreading drill cuttings). Additional description of this scenario is provided in NUREG–0782 (NRC, 1981, Appendix G).

5.1.3.3 Intruder-Recreational Hunting/Fishing Scenario

In this scenario, a hunter/fisher is assumed to visit the site, perhaps on a periodic basis, and to consume game and fish taken from the site, but otherwise not reside on the site. Fish and game ingestion pathways may need to be included for the intruder-resident or intruder-agriculture scenarios, but generally they result in limited relative contribution to total dose for these scenarios and can be eliminated from inclusion. The reviewer should assess the behavior attributed to the intruder and the timing of the intrusion. The reviewer should examine dose pathways assumed for this scenario (e.g., direct exposure, inadvertent soil ingestion, inhalation, ingestion of fish and game), the parameters and calculations used to estimate intruder doses, and the time the intruder is at the site.

5.1.3.4 Intruder-Driller Scenario

This scenario assumes that after the end of active institutional controls, a well is drilled into the waste disposal system. The well may be for domestic water use, irrigation, or the exploration or recovery of natural resources. If natural resources other than water are identified (see Section 1.1.3.2), additional drilling scenarios may be proposed. In a drilling scenario, an acute intruder is assumed to be the person (or persons) who installs the well and is exposed to drill cuttings during well installation. The reviewer should assess the behavior attributed to the intruder and the timing of the intrusion. The reviewer should examine dose pathways assumed for this scenario (e.g., direct exposure, inadvertent ingestion of soil, inhalation), the parameters and calculations used to estimate intruder doses, and the time spent to install the well. Additional description of this scenario is provided in NUREG/CR–4370 (Oztunali and Roles, 1986). Exposure of a resident or farmer to drill cuttings left on the land surface after the installation of a well would be considered under Section 5.1.3.1 or Section 5.1.3.2, respectively (this is referred to as a chronic scenario).

5.1.3.5   Intruder-Construction Scenario

In this scenario, it is assumed that after the end of active institutional controls, a construction project begins at the site with associated earthmoving activities, resulting in disruption of the disposal facility and potential exposure of waste to the environment. The reviewer should assess the behavior attributed to the intruder and the timing of the intrusion. The reviewer should examine dose pathways assumed for this scenario (e.g., direct exposure, inadvertent ingestion of soil, inhalation) and the time spent at the site. Additional description of this scenario is provided in NUREG–0782 (NRC, 1981, Appendix G).

5.1.3.6   Other Scenarios

There may be other less common scenarios proposed on a site-specific basis. As with the previous scenarios, the reviewer should assess the behavior attributed to the intruder and the timing of the intrusion. The reviewer should examine dose pathways assumed for the scenario, the time spent at the site, and the parameters and calculations used to estimate intruder doses. The reviewer should assess whether the other intruder scenarios are appropriate and sufficiently conservative.

5.2   Review Procedures

The reviewer should evaluate the types of scenarios DOE considered in the intruder analysis and confirm that the scenarios considered are appropriate for the site. Specifically, the reviewer should perform the following activities:

- Verify that assumptions and parameters used in defining the exposed intruder, including location and behavior of the intruder, timing of the intrusion, and exposure pathways, are consistent with the current regional practices. For example, at a site with shallow water resources and low-strength surface geologic materials, local drilling companies may not typically be equipped to drill through buried, high-strength engineered materials. However, at a site with high-strength geologic materials (e.g., basalt) and deep water resources, drilling companies may be commonly equipped to drill through an underground engineered structure (inadvertently).

- Verify that assumptions and parameters used in defining the exposed intruder are consistent with the assumptions and parameters used in the performance assessment that is reviewed using Section 4 of this guidance document. For example, confirm that the same radionuclide inventories are used for both the performance assessment and the intruder analysis.

- Assess whether, in the selection of intruder scenarios, DOE considered wasteform and barrier degradation (e.g., it may be safe to rule out drilling for the first 1,000 or 2,000 years as a result of the presence of intruder barriers or other wasteform characteristics, but drilling may become more plausible as the wasteform or barriers degrade).

- Verify that the time of intrusion assumed in the analysis produces the maximum dose. For example, an intrusion event at 100 years may produce the maximum dose from short-lived fission products, but the maximum dose from the daughters of long-lived isotopes may occur after 1,000 years.

- Verify that the wasteform properties and the disposal facility design (including natural and engineered barriers designed to protect inadvertent intruders) considered in the intruder analysis are consistent with the DOE-conducted performance assessment and are reviewed under Section 4 of this guidance document.

- Confirm that DOE does not use the probability of an intrusion to reduce the potential consequences estimated in the intruder analysis.

- Evaluate active institutional controls proposed in the waste determination, and assess the time period for which DOE assumes they are effective.

- Verify that proposed intruder barriers are appropriately represented in conceptual and numerical models used to simulate long-term performance. In particular, evaluate whether degradation of intruder protection systems is appropriately considered in the intruder analysis and that credit for the passive performance of intruder barriers is limited to 500 years or is otherwise justified.

- Verify that adequate technical basis is provided for parameters used in the intruder analysis, in particular for site-specific, regionally based values that are less conservative than nationally based generic parameters.

- Verify that the area over which contaminated material is dispersed is appropriate for the scenario. For example, the area required to distribute contaminated soil and waste from the excavation of a foundation for a residence will be considerably larger than the area required to distribute contaminated drill cuttings.

- If a garden is assumed in the scenario, verify that the garden size is appropriate and consistent with regional practices. Verify that the garden size is consistent with the assumed yields of produce from the garden.

- Evaluate whether the DOE assessment of inadvertent intrusion provides reasonable assurance that an inadvertent intruder will be sufficiently protected, based on an understanding of assumptions and parameters of the analysis, characteristics of the intrusion event, and consideration of uncertainties in the analysis. Compare the intruder doses calculated in the intruder analysis to the 5 mSv/yr [500 mrem/yr] standard described in NUREG–0872 (NRC 1981, Vol. 2, Section 4.5). Evaluate the DOE identification and summary of key assumptions and parameters that most strongly influence the dose results for the inadvertent intruder analyses.

5.3   References

Oztunali, O.I. and G.W. Roles. "Update of Part 61: Impacts Analysis Methodology, Methodology Report." NUREG/CR–4370. NRC. 1986.

U.S. Nuclear Regulatory Commission (NRC). "Draft Environmental Impact Statement on 10 CFR Part 61 Licensing Requirements for Land Disposal of Radioactive Waste." NUREG–0782. September 1981.

———. "Standard Review Plan for the Review of a License Application for a Low-Level Radioactive Waste Disposal Facility." NUREG–1200. NRC. January 1988.

———. "Savannah River Site High-Level Waste Tank Closure: Classification of Residual Waste as Incidental." Letter from W. Kane to R.J. Schepens, DOE. June 2000a.

———. "A Performance Assessment Methodology for Low-Level Radioactive Waste Disposal Facilities: Recommendations of NRC's Performance Assessment Working Group." NUREG–1573. October 2000b.

———. "NRC Review of Idaho Nuclear Technology and Engineering Center Draft Waste Incidental to Reprocessing Determination for Tank Farm Facility Residuals—Conclusions and Recommendations." Letter from L. Kokajko to J. Case, DOE. June 2003a.

———. "Consolidated NMSS Decommissioning Guidance." Vols. 1–3. NUREG–1757. September 2003b.

———. "Technical Evaluation Report for the U.S. Department of Energy, Savannah River Site Draft Section 3116 Waste Determination for Salt Waste Disposal." Letter from L. Camper to C. Anderson, DOE. December 2005.

6 PROTECTION OF INDIVIDUALS DURING OPERATIONS

The performance objective for protection of individuals during operations is provided in 10 CFR 61.43. Specifically, the staff review should confirm whether the design, operation, and closure of the facility will provide reasonable assurance that the radiation protection standards in 10 CFR Part 20 will be met. This includes doses both to workers and members of the public during operations. In addition, 10 CFR 61.43 requires that every reasonable effort will be made to maintain radiation exposures as low as is reasonably achievable (ALARA). The U.S. Department of Energy (DOE) is self-regulating with respect to operational activities and uses the regulations in 10 CFR Part 835, "Occupational Radiation Protection," to set operational dose limits for workers and members of the public and to demonstrate ALARA. The waste criteria discussed in Section 2 of this guidance document do not confer regulatory or statutory authority on the U.S. Nuclear Regulatory Commission (NRC) with regard to DOE operational activities.

In demonstrating that there is reasonable assurance that the 10 CFR Part 61.43 performance objectives will be met, DOE can reference the applicable portions of 10 CFR Part 20 and provide the corresponding requirement of 10 CFR Part 835. Given the role of NRC with respect to incidental waste activities, the portions of 10 CFR Part 20 that are relevant in this context are the dose limits for radiation protection of the public and workers during disposal operations and not those requirements regarding general licensing, administrative, programmatic, or enforcement matters intended for NRC licensees. Therefore, the applicable portions of 10 CFR Part 20 are

- 10 CFR 20.1101(d)
- 10 CFR 20.1201(a)(1)(i)
- 10 CFR 20.1201(a)(1)(ii)
- 10 CFR 20.1201(a)(2)(i)
- 10 CFR 20.1201(a)(2)(ii)
- 10 CFR 20.1201(e)

- 10 CFR 20.1206(e)
- 10 CFR 20.1207
- 10 CFR 20.1208(a)
- 10 CFR 20.1301(a)(1)
- 10 CFR 20.1301(a)(2)
- 10 CFR 20.1301(b)

These dose limits generally correspond to the dose limits in 10 CFR Part 835 and relevant DOE Orders and guidance that establishes DOE regulatory and contractual requirements for DOE facilities and activities. On the basis of this equivalence, the reviewer need only evaluate how the DOE regulations and limits would be implemented at the site. As the Ronald W. Reagan National Defense Authorization Act for Fiscal Year 2005 (NDAA) (see Appendix A) requires, NRC will monitor compliance with 10 CFR 61.43, as discussed in Section 10 of this guidance document, for waste that is subject to the NDAA.

6.1   Areas of Review

This part of the staff review focuses on the DOE-identified operational dose limits and the commitments made to meet them. Releases of radioactivity from facility effluents are governed by the requirements of 10 CFR 61.41, as discussed in Section 4 of this guidance document. Other doses are governed by 10 CFR 61.43, which references 10 CFR Part 20. The reviewer should determine what regulations and limits will be used to establish operational dose limits for any proposed facilities and activities and what approaches and radiation protection programs will be used to ensure that doses are maintained ALARA. The reviewer should evaluate estimated doses to both workers and members of the public during operations. Typically, the estimated dose to the public is calculated at the boundary of the facility where active institutional controls are not in effect (i.e., the fence line of the larger DOE site), but it may include the dose to members of the public who visit the site. As necessary, the reviewer should also examine selected portions of DOE operational documents to learn how DOE implements its worker protection program. For example, DOE may have radiation protection programs, documented safety analyses, or relevant access controls and training.

6.2   Review Procedures

It is anticipated that DOE will continue to reference the regulations and limits set forth in 10 CFR Part 835 to establish operational dose limits and to demonstrate that doses are maintained ALARA for proposed waste management activities. The implementation of these commitments and compliance with the requirements and limits will be ensured through DOE's own regulations and assessed during NRC monitoring (see Section 10). The review for this section should therefore focus on (1) DOE commitments to adhere to the appropriate regulations and (2) descriptions of how the regulations are implemented with respect to the waste determination. Based on this assumption, the reviewer should perform the following procedures:

- Determine whether DOE has identified either 10 CFR Part 20 or 10 CFR Part 835 as the basis for its radiation protection programs and dose limits. If DOE has identified other limits, ensure that these regulations and limits provide protection from radiation exposure that is comparable to the NRC requirements in 10 CFR Part 20 or 10 CFR Part 835.

- Evaluate the appropriateness and adequacy of DOE estimates of doses that may be received by both workers and members of the public during operations due to the waste being evaluated, and determine whether the doses are less than those in 10 CFR Part 20 or 10 CFR Part 835 {e.g., 50 mSv/yr [5 rem/yr] for workers, 1 mSv/yr [100 mrem/yr] to a member of the public from sources other than effluents}. Review the approaches and methods that assure the estimated doses will not be exceeded. The doses may be estimated by assessing the doses received from comparable activities at the site.

- Determine whether DOE has appropriate procedures and processes for ensuring that doses remain below applicable limits (e.g., a radiation protection program or safety analyses reports). Because DOE is a self-regulating Federal agency with a history of estimating and managing doses, the reviewer should apply the appropriate level of detail in reviewing this area (e.g., reviewing the assumptions made in accident analyses is typically not necessary).

- Verify that DOE will follow an ALARA philosophy. Confirm that management commitments, facility designs, and operations programs designed to control radiation exposure will be implemented to maintain doses ALARA. As necessary, examine selected portions of DOE operational documents to learn how DOE implements its worker protection program.

- Review any additional measures that DOE uses to ensure that radiation exposures will be ALARA.

7 SITE STABILITY, WASTE STABILITY, AND FACILITY STABILITY

7.1 Areas of Review

This section focuses on reviewing factors that could affect the stability of the proposed disposal site, including the potential effects of erosion, flooding, seismicity, and other disruptive processes. This section also addresses stability of the waste and engineered features of a disposal facility. The performance objectives for disposal site stability after closure are provided in 10 CFR 61.44, which states that the disposal facility must be sited, designed, used, operated, and closed to achieve long-term stability of the disposal site and to eliminate, to the extent practicable, the need for ongoing active maintenance of the disposal site following closure.

The long-term performance of the disposal site depends on the stability of the natural environment of the site, the disposal facility design, and the physical stability of waste disposed of at the facility. Disruptive events that are part of the natural environment have the potential to significantly degrade waste isolation by directly or indirectly affecting the engineered barriers or the wasteform. In general, disposal sites should not be susceptible to erosion, flooding, seismicity, or other disruptive events to such a degree or frequency that waste isolation is compromised. In addition to natural site instabilities, waste and disposal facilities also may be subject to instabilities because of waste characteristics (e.g., differential settling caused by voids in the waste) or facility design (e.g., long-term physical instability of vaults due to cracking). The relative importance of these processes may vary from site to site, and the level of detail of the review is expected to vary accordingly. Review areas are expected to include, but not necessarily be limited to, the technical areas described in 10 CFR 61.13(d) (i.e., erosion, mass wasting, slope failure, settlement of wastes and backfill, infiltration through covers over disposal areas and adjacent soils, and surface drainage of the disposal site).

7.1.1 Siting Considerations

The performance objective for site stability (10 CFR 61.44) requires, in part, that facilities be sited to achieve long-term stability. Although a waste determination may pertain to disposal in a facility that already has been sited (e.g., *in-situ* stabilization of residual waste in tanks), the siting considerations described in 10 CFR 61.50 should be reviewed because many of them identify processes that may affect long-term disposal site stability.[1] Processes that may affect site stability include, but are not necessarily limited to, flooding, runoff from upstream drainage, erosion, water table fluctuation, discharge to surface waters on the disposal site, subsurface geologic processes (e.g., seismic activity), slumping, landsliding, weathering, retrieval of subsurface natural resources, and activities at nearby facilities that could affect waste isolation. The reviewer should evaluate the consistency of the disposal site with the siting considerations. If there is significant potential for site instability, the reviewer should evaluate the measures taken to assure waste isolation despite the potential site instabilities.

---

[1]Another important purpose of the disposal site suitability requirements of 10 CFR 61.50 is to ensure that new waste sites are located in areas that can be modeled defensibly and monitored effectively. Therefore, some of the siting requirements, such as avoiding areas where nearby facilities or activities would significantly mask environmental monitoring programs, are not related to site stability.

7.1.1.1   Flooding and Water Table Fluctuation

Flooding, ponding, and periodic immersion of waste due to water table fluctuation all can affect the stability of the waste and disposal facility by accelerating wasteform and barrier degradation. To assess the potential for flooding and immersion of the waste due to water table fluctuation, the reviewer should evaluate site-specific hydrologic data, as described in Section 1.1.3.5. This includes evaluating rainfall intensity, the time of concentration of rainfall events, rainfall distributions, infiltration losses and surface runoff, and how these factors are considered in selecting probable maximum flood and the design basis flooding event. Estimates of the probable maximum flood and design basis flooding event should be consistent with climate projections that cover the full duration of the performance period, including uncertainties in those projections (see Section 4.3.1.1.3). Additional guidance on reviewing flooding information is provided in NUREG–1620 (NRC, 2003a, Section 3) and other U.S. Nuclear Regulatory Commission (NRC) guidance documents (NRC, 2005a, 2002).

To evaluate the potential for inundation of the waste by water table fluctuation, the reviewer should evaluate the historical record of water table depths in the area as well as information about the seasonal fluctuation of the depth of the water table. The reviewer should evaluate information about the wells from which the water table data were taken, including information about temporary surface features (e.g., paved areas on the surface near disposal units) that could artificially depress the water table elevation in wells near the proposed disposal site. In addition, if the waste is located near the zone of water table fluctuation, the reviewer should assess the potential rise in the water table due to potential increases in precipitation that could be caused by natural climate change during the 10,000-year performance period. As discussed in Section 4.3.1.1.3, the sensitivity of water table elevation to anthropogenic climate change can be assessed by assuming any projected change in the water table elevation due to climate change will occur sooner than expected due to natural climate cycling. If wastes may be located in the zone of water table fluctuation during the performance period, the reviewer should examine the technical basis for the predicted effect of the periodic inundation of the waste on the stability of the wasteform and relevant engineered barriers.

7.1.1.2   Surface Geologic Processes

Surface geologic processes such as erosion, mass wasting, slumping, and landsliding can have a significant effect on disposal site stability. Of the four sites discussed in this guidance, these processes are expected to have the greatest effect on the stability of disposal areas at the West Valley site. The reviewer should assess the potential for significant geological surface processes by reviewing historical information about such processes in the area (e.g., historical records of landslides near the site) and by reviewing information about site geomorphology as described in Section 1.1.3.4. The potential for surface geologic processes to affect site stability should account for short duration, large magnitude events (e.g., design basis rainfall events). The frequency and magnitude of these events should be consistent with climate states projected to occur during the performance period. Because there are other large sources of uncertainty in the prediction of surface geologic processes, including uncertainty resulting from present-day variability in rainfall events, predictions may not be sensitive to uncertainty in future climate states. If site stability is sensitive to uncertainty in future climate states, staff should incorporate current information regarding climate change into the review, as appropriate.

The reviewer should evaluate the design of engineered barriers proposed to provide long-term erosion protection. While active maintenance of erosion control systems may be assumed

1   during a period of institutional controls (100 years or less; see Section 4.1.1.1), longer term
2   erosion protection should rely on robust passive controls. To assess erosion control barriers,
3   the reviewer should consider rock durability; gradation; cover design; stability calculations for
4   the top slope, side slope, and apron for any cover; and other construction considerations that
5   are important to the performance of the erosion control system. Additional guidance about
6   reviewing the design of erosion control barriers is available in NUREG–1623 (NRC, 2002). As
7   described in NUREG–1623 and NUREG–1620 (NRC, 2003a, 2002), a review of proposed
8   erosion control measures also includes an assessment of the response of the engineered
9   system to flooding and precipitation events. Specifically, erosion barrier designs should account
10   for the selection of an appropriate design basis flood or rainfall event, control of gully initiation
11   and gully development, and the occurrence of flow concentrations and drainage network
12   development. As noted in Section 7.1.1.1, estimates of the design basis flood or rainfall event
13   should be consistent with climate projections that cover the full duration of the performance
14   period, including uncertainties in those projections.
15
16   **7.1.1.3   Subsurface Geologic Processes**
17
18   The reviewer should assess the potential for subsurface geologic processes, such as faulting or
19   seismic activity, to affect the site and proposed waste containment structures. For example,
20   although earthquakes are not frequent at the Savannah River Site, the Charleston earthquake
21   of 1886 had a magnitude estimate of M 7.3 with an epicenter approximately 144 km [90 mi] from
22   the Savannah River Site (NRC, 2005b). Such history indicates the need to evaluate the seismic
23   stability of the proposed waste containment structures and supporting soils; historic seismic
24   data should be reviewed in light of the most recent understandings of seismicity in the region
25   should be reviewed. The evaluation should include recurrence intervals, magnitudes, and
26   durations, as well as factors that contribute to peak ground acceleration such as underlying
27   geologic structures, and the stratigraphy and lithologies of the site. The predicted effects of
28   seismic events on waste isolation should be evaluated, and any aspects of the disposal plans
29   designed to mitigate the potential effects of seismic events on waste isolation should be
30   reviewed. Additional guidance on reviewing information related to seismic events is provided in
31   NUREG–1804 (NRC, 2003b).
32
33   Vulcanism is not expected to affect the stability of any of the sites discussed in this guidance.
34   Furthermore, the dose to an individual directly intruding into waste (see Section 5) is expected
35   to bound the probability-weighted potential doses due to vulcanism. For these reasons,
36   reviewers are not expected to assess the potential impacts of vulcanism unless new information
37   becomes available that indicates that (1) the potential for vulcanism at any of the relevant sites
38   is substantially greater than currently estimated and (2) the effects of vulcanism may not be
39   bounded by the dose to a potential inadvertent intruder.
40
41   **7.1.1.4   Other Processes**
42
43   Other natural processes may affect the long-term stability of the disposal site. These processes
44   can include biological processes such as biointrusion into waste or closure caps (e.g., by plant
45   roots or burrowing animals), weather-related hazards such as tornadoes and hurricanes, or
46   other hazards such as fires. During the operations period and the period of active institutional
47   controls, it is anticipated that DOE will have operational procedures (e.g., fire control) to limit the
48   effects of these processes. For the period following the end of institutional controls, these
49   processes should be considered on a case-by-case basis, depending on their potential to
50   disrupt the waste isolation capabilities of the disposal site. In many cases involving waste

1 buried at several meters depth, occasional surface events such as tornadoes and fires are not
2 expected to have a significant effect on disposal facility stability; one notable exception is for
3 evapotranspiration covers, because the plant community is susceptible to surface events such
4 as fires. The potential effects of near-surface processes on engineered barriers should be
5 bounded and, if necessary, evaluated in greater detail.
6
7 In addition to these natural processes, human activities also may affect site stability. For
8 example, the siting requirements of 10 CFR 61.50 indicate potential concerns associated with
9 retrieval of natural resources, activities at neighboring facilities, and population growth.
10 Reviewers should assume that DOE will limit the effects of anthropogenic disruptive processes
11 during the period of institutional controls (e.g., by limiting access to the site). To evaluate the
12 expected impact of human activities occurring after the institutional control period, the reviewer
13 should evaluate whether stabilizing features of the site will sufficiently limit the impact of human
14 activities on site stability. Because reprocessing waste typically includes long-lived
15 radionuclides, it would be unrealistic to attempt to predict population growth and future human
16 activities that may occur at the site during the entire period of interest (i.e., thousands of years).
17 Thus the reviewer should evaluate elements of the site designed to limit human intrusion
18 (e.g., disposal depth, recognizable wasteform, engineered intruder barriers) and determine
19 whether they provide reasonable assurance that human activities will not adversely affect site
20 stability during the performance period.
21
22 **7.1.2   Waste and Facility Stability**
23
24 In addition to the stability of the natural site, compliance with the performance objective for
25 disposal site stability depends on the stability of the waste and engineered barriers of the
26 disposal facility. Some of the concerns regarding the stability of low-level waste are related to
27 the stability of typical commercial Class A wastes, such as contaminated lab trash, clothing, or
28 plastics, and are not expected to be relevant to waste determinations. However, other aspects
29 of the stability of wastes described in 10 CFR 61.56(b), such as the structural stability of waste
30 under the overburden expected after site closure, the effect of radiation and changing chemical
31 conditions on the structural stability of the waste, the presence of free water in the waste, and
32 the presence of void spaces in the waste (e.g., in abandoned equipment), may be pertinent to
33 incidental wastes and should be reviewed. Similarly, the technical areas expected to affect
34 disposal site stability described in 10 CFR 61.51, including (1) the design of covers to limit water
35 infiltration, to direct water away from the waste, and to resist degradation by surface geologic
36 and biotic processes; (2) the design of surface features to direct surface water drainage away
37 from disposal units at velocities and gradients that will not result in erosion; and (3) the design
38 of the disposal site to minimize the contact of percolating or standing water with wastes after
39 disposal are all expected to be relevant to an assessment of disposal facility stability in the
40 context of a waste determination.
41
42 In addition to the stability concerns specifically described in 10 CFR Part 61, the reviewer
43 should also evaluate the potential for structural degradation of wasteforms and containment
44 structures (e.g., tanks, vaults) more specific to waste determinations (see Section 4.3.3). For
45 example, in many cases, waste may be mixed with or encapsulated by a cementitious material.
46 If so, the reviewer should evaluate the potential for structural degradation due to leaching,
47 sulfate attack, carbonation, corrosion of embedded metals, and cracking caused by various
48 mechanisms (e.g., shrinkage, differential settling, seismic activity) (see Section 4.3.2.2).
49 Additional guidance about evaluating the degradation of cementitious materials is available in
50 NUREG/CR–5542 (Walton, et al., 1990) and NUREG/CR–5666 (Clifton and Knab, 1989).

1   Degradation of the wasteform and containment facilities can increase groundwater penetration
2   into the waste and can provide enhanced paths for release of radionuclides. Typically, there
3   are large uncertainties associated with predictions of long-term wasteform degradation and
4   facility stability, and simplified analyses may bound the expected degradation (and the effects of
5   this degradation on performance) projected to occur during the performance period.
6
7   ## 7.2   Review Procedures
8
9   To evaluate compliance with the site stability performance objective (10 CFR 61.44), the
10   reviewer should assess the expected occurrence of the disruptive processes described in
11   Section 7.1 and evaluate the potential effects of the disruptive processes on disposal facility
12   performance. The level of detail of the review should depend on the potential for significant site
13   instability (e.g., flooding, erosion, seismicity) and the degree to which disruptive processes
14   could affect disposal facility performance. If the disposal site has a significant potential for
15   instability, the reviewer should perform a detailed review of the processes that could cause
16   instability and the elements of the proposed facility designed to ensure waste isolation in light of
17   the potential instabilities. If, on the other hand, there is little potential for significant disposal site
18   instability, then the reviewer should conduct a simplified review focusing on the technical bases
19   for this conclusion.
20
21   This part of the review is limited to an assessment of site stability. However, processes that
22   cause significant disposal site instability are expected to affect the long-term performance of the
23   disposal site. For example, increased infiltration due to disruption of a closure cap by
24   biointrusion or erosion may have significant effect on radionuclide release and may need to
25   be considered in the performance assessment (see Section 4). Similarly, structural
26   degradation of wasteforms or intruder barriers would be expected to have a significant effect
27   on the plausibility of potential inadvertent intruder scenarios (see Section 5). Thus, the reviewer
28   should coordinate the review of site stability with the review of compliance with other
29   performance objectives of 10 CFR Part 61, Subpart C. The reviewer should perform the
30   following procedures:
31
32   •   Confirm that the evaluation of flooding scenarios has accounted for flooding of adjacent
33   streams, as applicable, and localized flooding of drainage channels and protective
34   features. Verify that DOE has properly used the probable maximum
35   precipitation/probable maximum flood in determining the design flood event (NRC, 2002,
36   Appendix D; 2003a, Section 3.2.2).
37
38   •   Confirm that the probable maximum flood is consistent with, but not based solely on an
39   extrapolation of the historic flood record. Ensure that the technical basis for the
40   probable maximum flood includes calculations based on the most severe reasonably
41   possible rainfall events that could occur as a result of a combination of the most severe
42   meteorological conditions occurring over a watershed (probable maximum precipitation).
43
44   •   Verify that estimates of the probable maximum precipitation and probable maximum
45   flood are consistent with climate states that are expected to occur during the
46   performance period. Verify that the effects of uncertainties in projected climate states
47   are appropriately incorporated into or bounded by estimates of the probable maximum
48   precipitation and probable maximum flood.
49

- If the historic maximum regional floods exceed or closely approximate the proposed probable maximum flood estimates, perform a detailed evaluation to determine the basis for the estimates. Compare basin lag times, rainfall distributions, soil types, and infiltration loss rates to determine whether there is a logical basis for the probable maximum flood values being less than historic floods.

- If DOE uses detailed computer models to support its design flood determinations, confirm the adequacy of the various input parameters to the model, including, but not limited to drainage area, lag times and times of concentration, design rainfall, incremental rainfall amounts, temporal distribution of incremental rainfall, and runoff/infiltration relationships.

- Determine whether the waste is likely to be located in the zone of water table fluctuation during the 10,000-year performance period. Specifically, the reviewer should:

  - Evaluate the historical record of water table depths in the area as well as information about seasonal fluctuations of water table depths. The reviewer should examine precipitation data corresponding to the period of historical water table data to determine whether the water table data were taken during an adequately representative period of precipitation (e.g., that well data do not represent a period of relatively low precipitation).

  - Ensure that the assessment of the waste location with respect to the water table is not affected by temporary surface features (e.g., paved areas on the surface near disposal units) that could artificially depress the water table elevation in wells near the proposed disposal site.

  - Determine whether the waste may be located in the zone of water table fluctuation during the 10,000-year performance period because of potential changes in precipitation due to climate change. Note that either a rise or fall in the water table (due to wetter or drier future climates) could cause the waste to be located in the zone of water table fluctuation, depending on whether the waste is currently located above or below the water table (e.g., some tanks in H tank farm at SRS are partially submerged). Evaluate the potential impacts of anthropogenic climate change by assuming that changes in water table elevation may occur sooner than would be predicted based on natural climate cycling.

  - If the waste may be located in the zone of water table fluctuation during the performance period, examine the technical basis for the predicted stability of the disposal site and wasteform in detail, and ensure that the potential effects of water table fluctuation on disposal site stability have been adequately represented or bounded (e.g., effects on wasteform degradation, barrier degradation, sudden radionuclide releases due to episodic submersion of waste).

- Verify that the analysis of potential erosion at the site includes an assessment of floods, flood velocities, design features, and rock durability that is comparable to those described in NUREG–1623 (NRC, 2002, Appendix D).

- Confirm that estimates of the potential for surface geologic processes such as erosion, slumping, slope failure, and landsliding are consistent with historical site information (e.g., information about landslides at the site) and site geomorphology.

- Verify that estimates of the potential effects of surface geologic processes, including erosion, slumping, slope failure, and landsliding, account for short-duration, large magnitude events (e.g., design basis rainfall events).

- Verify that the analysis of potential erosion, slumping, slope failure, and landsliding are consistent with projections of climactic conditions that cover the duration of the performance period, including potential changes in the frequency or magnitude of short-duration, large magnitude events.

- Evaluate the sensitivity of site stability to the timing of the effects of climate change (e.g., changes in precipitation, infiltration, flooding, soil saturation, water table depth, erosion, and landsliding) as a means to evaluate the potential effects of anthropogenic processes affecting climate change.

- If the estimated impacts of flooding, water table fluctuation, or surface geologic processes are sensitive to uncertainty in climate change, consider bounding assumptions based on current information about climate change, as appropriate.

- If erosion is expected to have a significant effect on site stability, evaluate whether the erosion protection design is sufficient to avoid the need for ongoing active maintenance at the site. Specifically, determine the adequacy of the following technical areas, as appropriate to the proposed design:

  — Treatment of the banks of natural channels, including armoring and designs to place riprap as a protective measure against flood erosion;

  — Stability estimates for the top slope, side slopes, and apron, including design flow rate, depth of flow, design discharge, angle of repose, specific gravity, rock sizes, and other parameters;

  — The design of diversion channels, outlets, and discharge areas, including the parameters used to define the erosion protection such as flow rates, flow depths, shear stresses, erosion protection (riprap), rock size, and the effects of sediment accumulation;

  — Rock durability testing of proposed rock sources, especially with regard to clay content, to ensure that durable rock will be used;

  — Determination of allowable shear stresses and permissible velocities for any soil or vegetative cover, including an assessment of the cover performance in a degraded state;

  — Information on types of vegetation proposed and their abilities to survive natural phenomena and expected natural vegetation progression (e.g., grass cover being replaced by tree cover over time);

— Construction considerations such as plans, specifications, inspection programs, and quality assurance/quality control programs to assure that adequate measures are being taken to construct the design features according to accepted engineering practices; and

— Information, analyses, and calculations of input parameters to models used.

- Verify that the description of the potential for seismic events is derived from the historical record, paleoseismic studies, or geological analyses. Evaluate recurrence intervals and magnitudes of past events, and determine whether the reported potential for seismic events is consistent with the historical record.

- Determine whether seismic events that are likely to occur during the performance period are likely to have a significant impact on the stability of the disposal facility by causing structural damage to wasteforms or engineered barriers (e.g., cracking of grouted wasteforms or vaults).

- Assess the potential disruption of wasteforms or engineered barriers (e.g., caps designed to limit infiltration) by intrusion of roots or burrowing animals by evaluating biological information relevant to the site. If populations of burrowing animals are identified or if deep-rooted plant species are expected to grow on the site (e.g., if the site is expected to revert to forests) during the performance period, determine whether disposal site features designed to limit the effects of biointrusion are adequate to maintain site stability.

- Determine whether there are natural resources at the site such that exploitation of the natural resources (e.g., mining for mineral ore) would affect site stability. If so, determine whether features of the wasteform or disposal site (e.g., a recognizable wasteform or engineered barriers) would be sufficient to alert individuals to the presence of waste or otherwise limit their impact on site stability (see review procedures related to engineered barriers and wasteform degradation in Sections 4.3.2 and 4.3.3).

- Determine whether current or expected activities at nearby facilities (e.g., use or decommissioning of neighboring chemical processing facilities on the DOE site) may affect site stability. If so, determine whether disposal site features designed to limit the effects of the activities are sufficiently robust to provide reasonable assurance that site stability will be maintained (see review procedures related to engineered barriers and wasteform degradation in Sections 4.3.2 and 4.3.3).

- If there are significant quantities of long-lived radionuclides at the site, determine whether wasteforms are likely to remain sufficiently identifiable and intruder barriers are likely to remain sufficiently robust to provide reasonable assurance that human activities will not compromise site stability thousands of years after site closure (see review procedures related to engineered barriers and wasteform degradation in Sections 4.3.2 and 4.3.3).

- Determine whether disruptive events such as hurricanes, tornadoes, and fires could cause significant site instability, and if so, what aspects of the facility are designed to maintain site stability if the disruptive events occur.

1 • Verify that the waste will remain structurally stable under the overburden expected after
2   site closure.
3
4 • Verify that the effects of radiation and any changes expected in chemical conditions
5   (e.g., because of infiltrating groundwater or corrosion of embedded metal) will not cause
6   instability of the waste as emplaced after closure.
7
8 • Verify that the presence of free water in the waste will be minimal (e.g., less than
9   1 percent by volume) and that there will not be significant void spaces in the waste
10   (e.g., inside abandoned equipment) that could cause differential settling of the waste or
11   engineered barriers.
12
13 • Verify that covers have been designed to limit water infiltration and to direct infiltrating
14   water away from the waste. Verify that any surface features of the disposal facility have
15   been designed to direct surface water drainage away from disposal units at velocities
16   and gradients that will not result in erosion.
17
18 • If wastes will be mixed with or encapsulated by cementitious material, determine the
19   potential for significant structural degradation due to leaching, sulfate attack,
20   carbonation, or corrosion of embedded metals. Because of the large uncertainty in
21   potential degradation through these processes, use simplified analyses to bound the
22   expected degradation during the performance period. If there is significant potential for
23   structural degradation of the wasteform or cementitious engineered barriers because of
24   these processes, the potential effects on radionuclides release or plausible inadvertent
25   intruder scenarios should be represented or bounded in the performance assessment
26   (see Section 3) or inadvertent intruder analysis (see Section 4).
27

7.3   References

Clifton, J.R. and L.I. Knab, "Service Life of Concrete." Final Report. NUREG/CR–5466.
September 1989.

U.S. Nuclear Regulatory Commission (NRC). "Design of Erosion Protection for Long-Term
Stabilization." Final Report. NUREG–1623. September 2002.

———. "Standard Review Plan for the Review of a Reclamation Plan for Mill Tailings Sites
Under Title II of the Uranium Mill Tailings Radiation Control Act of 1978." NUREG–1620. June
2003a.

———. "Yucca Mountain Review Plan." NUREG–1804. July 2003b.

———. "Consolidated NMSS Decommissioning Guidance. Updates to Implement the License
Termination Rule Analysis." Draft Report for Comment. NUREG–1757, Supplement 1.
September 2005a.

———. "Final Safety Evaluation Report on the Construction Authorization for the Mixed Oxide
Fuel Fabrication Facility at the Savannah River Site, South Carolina." NUREG–1821. March
2005b.

1 Walton, J.C., L.E., Plansky, and R.W. Smith. "Models for Estimation of Service Life of Concrete
2 Barriers in Low-Level Radioactive Waste Disposal." Final Report. NUREG/CR–5542.
3 September 1990.

8 QUALITY ASSURANCE PROGRAM

Quality assurance, in the context of this guidance document, comprises all of the planned and systematic actions necessary to provide adequate confidence that the applicable incidental waste criteria will be met. An adequate quality assurance program is essential to ensuring that the key information relied upon to make incidental waste determinations is correct and accurate.

This review verifies that the U.S. Department of Energy (DOE) has applied quality assurance measures to its data collection, analyses, waste determinations, and performance assessments. This guidance document provides risk-informed and performance-based approaches for evaluating quality assurance information. The adequacy of quality assurance measures will be determined using a sample of analyses selected based on their risk significance, and conclusions will be based on impacts of any identified deficiencies on performance. For example, a quality assurance deficiency of a risk-significant element of the performance assessment may have a minor, moderate, or major impact on the results of the analysis. The reviewer should consider both the severity and the pervasiveness of deficiencies when evaluating the adequacy of quality assurance measures. The reviewer should use this guidance in reviewing quality assurance measures that apply to DOE waste determinations and performance assessments. Section 17.6 of NUREG–1757, Volume 1, Revision 1, provides U.S. Nuclear Regulatory Commission (NRC) regulatory guidance pertinent to quality assurance as applied to decommissioning sites (NRC, 2003). The reviewer may consider this guidance when evaluating DOE's quality assurance program as applied to the waste determination.

8.1   Areas of Review

The NRC staff should evaluate a sample of the analyses associated with Areas of Review identified in Sections 2–7 of this guidance document. The analyses should be selected based on high or medium risk areas with regard to meeting the performance objectives of 10 CFR Part 61, Subpart C, and other applicable criteria. The risk importance of an analysis is likely to vary from facility to facility.

8.2   Review Procedures

8.2.1   Data Validity Review Procedures

The reviewer should examine the data used to support waste determinations and performance assessments to determine whether (1) the data are traceable to their sources through all calculations and data reductions, and the data are transparent in their use and (2) the data have been obtained or qualified under an acceptable quality assurance program, such as but not limited to an NRC-approved quality assurance program developed to meet the requirements of 10 CFR Part 50, Appendix B, or are otherwise documented and validated. Specifically, the reviewer should ensure the following conditions are met:

- Data are identified in a manner that facilitates traceability to associated documentation back to the source and clearly identifies qualification status (e.g., qualified, unqualified, or accepted, such as for a physical constant). Traceability is the ability to trace the history, application, or location of an item and like items or activities through recorded identification.

- Any changes that affect data identification (e.g., category or use/application) are made in a manner that preserves traceability.

- Data used as direct input to scientific analysis or performance modeling are qualified or are validated by other comparable methods.

- If data are not collected under an acceptable quality assurance program, the data are qualified by one or more of the following processes: equivalent quality assurance program, comparison to corroborating data, confirmatory testing, or peer review (NRC, 1988).

- Documentation regarding data traceability and qualification is transparent and identifies the principal lines of investigation considered. A document is transparent if it is sufficiently detailed as to purpose, method, assumptions, inputs, conclusions, references, and units so that a person technically qualified in the subject can understand the document and ensure its adequacy without recourse to the originator.

- The data reduction process is described in detail sufficient to allow independent reproducibility by another qualified individual. Data reduction includes processes that change the form of expression, quantity of data or values, or number of data items. Verify that data reduction inputs, outputs, and computational methods are documented.

8.2.2   Software Selection and Development Review Procedures

The reviewer should assess the software DOE used in the waste determination and performance assessment to determine whether (1) the software has been endorsed for NRC use or (2) DOE has demonstrated that the software adequately represents the processes or systems for which it is intended. The reviewer should evaluate whether the software development and approaches to software validation are planned, controlled, and documented. Planning for validation identifies the validation methods and validation criteria used.

The reviewer should assess controls applied to software to ensure that the software supporting waste determinations or performance assessments is qualified for use and has been developed, tested, and controlled under suitable conditions.

Specifically, the reviewer should ensure the following conditions are met:

- NRC has endorsed the software for use in licensing activities. If the software has been endorsed, then the reviewer would, in general, not need to exercise the review procedures specific to software quality assurance that follow.

- The software performs intended functions, provides correct solutions, and does not cause adverse unintended results.

- Software verification and validation activities were planned, documented, and performed for each item of software.

- Software that was verified and validated and was subsequently changed has undergone additional verification and validation, and documentation of the additional verification and validation has been developed.

- The software development and maintenance process proceeded in a planned, traceable, and orderly manner, using a defined software life-cycle methodology.

- A software configuration management system has been established.

- Requirements controlling software procurement and services are established to assure proper verification and validation support, software maintenance, configuration control, and performance of software audits, assessments, or surveys. Requirements for suppliers reporting software errors to the purchaser and, as appropriate, the purchasers reporting software errors to the supplier are identified.

- If a defect was identified in software that adversely affects the results of previous software application, the condition adverse to quality was documented and controlled.

8.2.3   Analysis Review Procedures

The reviewer should assess the analysis to determine that the analysis was transparently documented, the objective and use of the software are described, and the software was not used outside of the range of its intended functions. Specifically, the reviewer should ensure that the following conditions are met:

- The definition of the objective (intended use) of the software has been provided;

- The description of the conceptual model implemented in the software is clear;

- The software has not been used in a manner that is inconsistent with the conceptual model implemented in the software;

- Identification of inputs to the software and their sources has been provided;

- Discussion of mathematical and numerical models that are used in the software is provided, including governing equations, formulas, algorithms, and their scientific and mathematical bases;

- Associated software used, computer calculations performed, and basis to permit traceability of inputs and outputs have been identified;

- Discussion of initial and boundary conditions is provided;

- Model limitations (e.g., data available for model development, valid ranges of model application, spatial and temporal scaling) are discussed; and,

- Software execution is appropriate with the various sources of uncertainties (i.e., conceptual model, mathematical model, process model, system model, parameters).

8.3 References

U.S. Nuclear Regulatory Commission (NRC). "Generic Technical Position on Qualification of Existing Data for High-Level Nuclear Waste Repositories." NUREG–1298. Washington, DC. February 1988.

————. "Consolidated NMSS Decommissioning Guidance." Vols. 1–3. NUREG–1757. Washington, DC. September 2003.

9 DOCUMENTING THE RESULTS OF THE REVIEW

9.1 General Approach to Documenting Waste Determination Reviews

As described in the U.S. Nuclear Regulatory Commission's (NRC) implementation plan for waste determination reviews conducted under the Ronald W. Reagan National Defense Authorization Act for Fiscal Year 2005 (NDAA) (NRC, 2005a), the general approach the NRC staff will use will be similar to previously completed waste-incidental-to-reprocessing (WIR) reviews (NRC, 2000, 2002, 2003). Reviews not conducted under the NDAA will be conducted in a similar manner. As discussed in Section 2, however, due to differences between the criteria in the NDAA and those used in previous NRC reviews, there may be some technical differences in the reviews. This section of the guidance document provides guidance on the process for conducting and documenting a review.

9.2 Request for Additional Information

The first step in the review process will be the U.S. Department of Energy's (DOE) submittal of a draft waste determination and supporting documentation, including a performance assessment if necessary. Using the guidance described in Sections 1–8 of this document, the NRC staff will review whether DOE's assumptions, analyses, data, documentation, modeling, and conclusions are technically adequate, accurate, and in accordance with the appropriate waste criteria. If the reviewer does not receive sufficient information to determine that there is reasonable assurance that the waste criteria can be met or if the reviewer has questions about the DOE-provided information, the NRC staff should develop a Request for Additional Information (RAI). An RAI is a list of questions needing DOE responses so NRC staff can complete its review. The staff's RAI should be risk-informed and focus on those areas that are most likely to affect the staff's conclusions. The RAI should be as complete as possible so that only one RAI is needed. In addition to responding to the specific questions raised in the RAI, DOE may decide to revise the waste determination itself, or the supporting documentation or modeling, based on NRC's questions and comments. If the information DOE provided in its initial submittal is sufficient for the reviewer to determine whether there is reasonable assurance that the applicable waste criteria can be met, then the staff does not necessarily need to prepare an RAI and can prepare its Technical Evaluation Report (TER) (see Section 9.3). NRC's role in evaluating waste determinations is consultative; as such, DOE has final authority to determine whether the applicable waste criteria can be met.

9.3 Technical Evaluation Report

The TER documents the NRC staff's analyses and conclusions for a specific waste determination. The TER should include descriptions of DOE's approach, what was reviewed by the staff, the assumptions made in conducting the review, the major sources of uncertainty in DOE's analysis, and the conclusions as to whether there is reasonable assurance that each applicable waste criterion can be met (see Sections 2–7). The amount of discussion in the TER for a specific area should be commensurate with its importance to NRC's conclusions. Examples of areas typically covered in a TER are waste inventory, identification of highly radioactive radionuclides, infiltration, wasteform degradation, near field transport, and hydrology. Specific TER sections may be developed as necessary to evaluate those aspects that are most significant for a specific waste determination. The TER may also include, in an

1 appendix, recommendations for DOE's consideration; the purpose of the recommendations is to
2 communicate actions that DOE might consider to further improve its waste management
3 approach. These do not need to be implemented for the applicable waste criteria to be met.
4
5 For waste determination reviews conducted under the Ronald W. Reagan National Defense
6 Authorization Act for Fiscal Year 2005 (NDAA) (see Section 2), the TER should also identify the
7 key aspects that are important to assessing compliance with 10 CFR Part 61, Subpart C. These
8 key aspects will be one facet of the NRC's monitoring role under the NDAA for a particular
9 waste determination (see Section 10). Ideally, the detailed technical evaluation will have been
10 completed during consultation with DOE and prior to the initiation of NRC's monitoring activities
11 which begin when DOE issues the final waste determination document. NRC staff should
12 review DOE's final waste determination document and evaluate any differences between it and
13 the draft waste determination document.
14

15 9.4   Public Availability and Project Numbers

16
17 The Commission has directed the staff to ensure that the technical basis for its decisions
18 regarding draft waste determination reviews under the NDAA are as "transparent, traceable,
19 complete, and as open to the public and interested stakeholders as possible" (NRC, 2005b). To
20 fulfill that direction, the documents associated with DOE's waste determinations and NRC's
21 reviews should be made publicly available, both for reviews that are being conducted for sites
22 under the NDAA and those that are not. This includes the waste determinations, supporting
23 references, NRC's RAI, DOE's RAI responses and supporting references, meeting summaries,
24 TERs, and any other relevant documents submitted by DOE or issued by NRC. One exception
25 may be documents that DOE cannot publicly release because of security concerns. For
26 discussion of public availability of reports related to monitoring under the NDAA, see Section 10.
27
28 For ease of finding and obtaining documents, the NRC staff has established project numbers for
29 incidental waste activities at the sites, as follows: Savannah River Site is PROJ0734, Idaho
30 National Laboratory is PROJ0735, Hanford is PROJ0736, and West Valley is POOM–32.
31 These project numbers should be entered into the "Docket Number" field in NRC's Agencywide
32 Documents Access and Management System (ADAMS).
33

34 9.5   References

35
36 U.S. Nuclear Regulatory Commission (NRC). "Savannah River Site High-Level Waste Tank
37 Closure: Classification of Residual Waste as Incidental." Letter from W. Kane to
38 R.J. Schepens, DOE. June 2000.
39
40 ——. "NRC Review of Idaho National Engineering and Environmental Laboratory Draft Waste
41 Incidental to Reprocessing Determination for Sodium-Bearing Waste—Conclusions and
42 Recommendations." Letter from J. Greeves to J. Case, DOE. August 2002.
43
44 ——. "NRC Review of Idaho Nuclear Technology and Engineering Center Draft Waste
45 Incidental to Reprocessing Determination for Tank Farm Facility Residuals—Conclusions and
46 Recommendations." Letter from L. Kokajko to J. Case, DOE. June 2003.
47

1   ———. "Implementation of New U.S. Nuclear Regulatory Commission Responsibilities Under
2   the National Defense Authorization Act of 2005 in Reviewing Waste Determinations for the
3   U.S. Department of Energy." SECY–05–0073. April 2005a.
4
5   ———. "Staff Requirements—SECY–05–0073—Implementation of New U.S. Nuclear
6   Regulatory Commission Responsibilities Under the National Defense Authorization Act of 2005
7   in Reviewing Waste Determinations for the USDOE." SRM–SECY–05–0073. June 2005b.
8

10 NDAA COMPLIANCE MONITORING

Paragraph (b)(1) of Section 3116 of the Ronald W. Reagan National Defense Authorization Act for Fiscal Year 2005 (NDAA) (see Appendix A) requires that the U.S. Nuclear Regulatory Commission (NRC) " ... in coordination with the covered State, monitor disposal actions taken by the Department of Energy ... for the purpose of assessing compliance with the performance objectives set out in subpart C of part 61 of title 10, Code of Federal Regulations." The NDAA requires that NRC report any noncompliance to Congress, the State, and the U.S. Department of Energy (DOE) as soon as practicable after discovery of the noncompliant conditions and states that NRC's monitoring is subject to judicial review. However, the NRC does not have regulatory or enforcement authority over DOE. The NDAA applies only to South Carolina and Idaho, and these are the States in which NRC would monitor DOE's disposal of incidental waste. NRC does not have a monitoring role at Hanford, because Washington is not included as a covered State in the NDAA.

For West Valley, the NRC license is currently in abeyance while DOE completes its responsibilities under the West Valley Demonstration Project Act (WVDPA). The WVDPA requires that NRC establish decommissioning criteria for the site. NRC published those criteria in 2002 (NRC, 2002) and included criteria for determining whether certain waste disposed of onsite is incidental waste; however, NRC does not have regulatory or enforcement authority over DOE under the WVDPA. Although the WVDPA does assign a monitoring role to NRC at West Valley, at this point, NRC staff has not performed a technical evaluation of a waste determination at WVDP to assess compliance with safety requirements comparable to the performance objectives in 10 CFR Part 61, Subpart C. Therefore, the West Valley site is not included in the following discussion of monitoring. Monitoring approaches and activities relating to incidental waste disposal actions for WVDP can be included in a revised and updated version of the guidance document when relevant. If the license is reinstated and responsibility for the entire site returns to the licensee (the New York State Energy Research and Development Authority [NYSERDA]), NRC will retain the same monitoring and inspection responsibilities that it has for other licensees under the Atomic Energy Act with respect to ensuring that the site meets all applicable regulatory requirements.

10.1   Overall Approach and Scope

The NDAA requires NRC to monitor DOE's disposal actions to assess compliance with the performance objectives in 10 CFR Part 61, Subpart C. The NDAA also directs the NRC to issue a report to Congress, the covered State, and DOE if it determines that disposal actions DOE takes are not in compliance with the performance objectives. The act does not provide NRC with any regulatory or enforcement authority. Accordingly, NRC monitoring activities will be limited to those activities deemed necessary to assess whether there is reasonable assurance that compliance with the performance objectives will be achieved.

As stated in SECY–05–0073 (NRC, 2005), the staff intends to carry out its monitoring role in a risk-informed and performance-based manner. A cornerstone of the staff's intended approach is to identify key aspects of DOE's disposal actions that should be monitored. Identifying these key aspects is expected to be part of a technical evaluation to assess whether DOE's proposed waste disposal actions will comply with 10 CFR Part 61, Subpart C, performance objectives.

1   To the extent possible, the staff will conduct its technical evaluation during the consultation
2   process and leverage insights gained through the consultation process to identify key aspects of
3   DOE's waste disposal system that should be monitored to assess compliance with NRC's
4   10 CFR Part 61 performance objectives. Staff intends to use risk insights gained through its
5   technical evaluation to gauge the relative importance of key modeling assumptions or model
6   parameters DOE uses to demonstrate performance objectives will be met. Assumptions,
7   parameters, and features that are expected to have a large influence on the performance
8   demonstration and/or have relatively large uncertainties will be considered key aspects of the
9   waste disposal system. Monitoring is a good mechanism to manage those uncertainties, and
10 NRC staff should monitor the disposal actions with which these key assumptions, parameters,
11 and features are associated. Monitoring is also a way to evaluate new information that may
12 potentially support previous information or assumptions. In either case, the performance
13 assessment in DOE's final waste determination document to demonstrate compliance with the
14 performance objectives must continue to be adequately supported. System description, data
15 sufficiency, data uncertainty, model uncertainty, and model support ensure the performance
16 assessment is technically adequate; these are the main elements in the general technical
17 review procedures as described in the early part of this guidance document. NRC staff also
18 recognizes that some uncertainties will exist at the time that decisions are to be made.
19 Although it is not practicable to require a performance assessment to be assumption free, it is
20 reasonable to expect that key assumptions have a technical basis supported by data and
21 information. Monitoring is not to be used as a substitute for inadequate information, but rather
22 to manage the uncertainty and to support the previous determination of adequacy considering
23 uncertainty. Focusing on the most risk-significant aspects of the performance demonstration
24 will allow staff to optimize its resources.
25
26 The NDAA indicates NRC must monitor DOE's disposal actions to determine whether the
27 actions are in compliance with the performance objectives of 10 CFR Part 61, Subpart C.
28 Because the performance objectives include a general requirement that there is reasonable
29 assurance that exposures to humans will not exceed the limits established in 10 CFR 61.41
30 through 44, DOE's disposal actions would be out of compliance if NRC no longer had
31 reasonable assurance that the requirements of 10 CFR 61.41 through 44 would be met. As
32 described in Sections 4–7 of this guidance document, NRC staff typically derives assurance that
33 the requirements of 10 CFR 61.41 through 44 will be met on predictions of long-term site
34 performance. Therefore, monitoring to assess compliance with the performance objectives is
35 expected to include activities necessary to maintain confidence in DOE's predictions of
36 long-term site performance. Although environmental monitoring (i.e., monitoring the
37 environment for changes and collecting environmental data) is expected to provide the most
38 clear and straightforward way to assess the performance of a disposal facility, in practice, there
39 are limitations to solely relying on environmental data to evaluate the performance of the type of
40 disposal facilities that are the subject of DOE's waste determinations. Because DOE typically
41 relies on a number of engineered features (e.g., grout, concrete vaults, specialized covers) to
42 close their facilities, it may be several decades or centuries before any radioactive materials are
43 expected to be released from the disposal facility and could thereafter be detected. While
44 observing DOE actions and reviewing environmental data are expected to be components of
45 NRC's monitoring activities, building confidence in DOE's selection of parameters and models
46 will be a critical monitoring activity. Accordingly, it is important for staff to use the performance
47 assessment to identify those aspects of the disposal system that have the largest influence on
48 performance so that these areas can be the focus of NRC's monitoring activities. In addition, it
49 is important for staff to monitor those parts of the disposal system with the greatest amount of
50 uncertainty that NRC's performance objectives will be met.

The staff's approach to assessing compliance with the performance objectives will generally consist of two primary activities: (1) conducting technical reviews of DOE data and analyses and (2) physically observing DOE's disposal actions through onsite visits. Table 10-1 shows the expected primary monitoring activities for assessing compliance with each of the performance objectives.

Based on the NRC staff's technical evaluation, NRC may conclude that a number of key aspects associated with DOE's performance assessment should be monitored to assess compliance with the performance objectives. For example, DOE may be able to demonstrate that the performance objectives will be met based upon a demonstrably conservative analysis, in which case, less uncertainty would exist regarding assumptions and parameters important to meeting the performance objectives. On the other hand, the staff's technical evaluation may

Table 10-1. U.S. Nuclear Regulatory Commission's Primary Monitoring Activities for Each of the Performance Objectives	
Performance Objective	**Primary Monitoring Activities**
10 CFR 61.40—General Requirements	Addressed through compliance with 10 CFR 61.41–44
10 CFR 61.41—Protection of General Population	1. Review information and observe actions relevant to key aspects important to demonstrating compliance with 10 CFR 61.41 2. Review DOE environmental monitoring results 3. Review DOE sampling data and inventory estimates 4. Observe environmental sampling and data collection activities 5. Review updates to DOE's performance assessment
10 CFR 61.42—Protection of Inadvertent Intruder	1. Review information and observe actions relevant to key aspects important to demonstrating compliance with 10 CFR 61.42 2. Review intruder barrier and erosion control designs that can be used to prevent intruder scenarios 3. Similar compliance monitoring activities that are affiliated with 10 CFR 61.41
10 CFR 61.43—Protection of Individuals During Operations	1. Review DOE worker and public radiation records 2. Review DOE and covered state environmental monitoring data, offsite dose calculations, and radiological assessments
10 CFR 61.44—Site Stability	1. Observe construction of engineering features and their maintenance 2. Observe for signs of structural failure 3. Review design and design changes of the engineered features

1   conclude that there are key aspects of the performance assessment where a significant amount
2   of uncertainty exists and continues to exist during the decision making process. In either case,
3   staff should remain aware of the implementation of DOE's waste management approach and
4   any developments that may challenge the support of DOE's assumptions or analyses. The key
5   aspects listed in the TER are not intended as an all-inclusive list of monitoring activities, and the
6   number or types of monitored activities may change as more is learned about the disposal
7   methods or as DOE's disposal plans proceed. NRC staff performing the monitoring should
8   evaluate revisions to DOE's modeling or disposal plans.
9
10   Once NRC has completed its technical evaluation and DOE has issued a final waste
11   determination, NRC staff should have completed or be in the process of completing a
12   compliance monitoring plan for carrying out its specific monitoring activities. Ideally, NRC will
13   develop and be prepared to implement its monitoring plan before DOE initiates waste disposal
14   activities. Consultations with DOE should ensure, to the extent practical, that NRC's
15   compliance monitoring plan is in place before waste determination activities begin. Waste
16   removal or retrieval may be the exception; this disposal action is different from other waste
17   determination activities in that waste retrieval may be in progress when the final waste
18   determination has been issued, or waste retrieval activities may be completed. NRC staff will
19   monitor waste retrieval activities after DOE issues the final waste determination. NRC's
20   compliance monitoring plan will identify the specific activities that will be undertaken to assess
21   DOE's compliance with each of the performance objectives. Specific elements of the monitoring
22   plan will be established based upon the individual waste determination. However, some
23   elements are expected to be common to most monitoring plans and important to waste isolation,
24   including (1) reviewing the final disposal facility inventory of radionuclides, (2) reviewing updates
25   to DOE's performance assessment, (3) observing the construction and maintenance of
26   engineered systems (e.g., vaults and covers), (4) reviewing radiation records, (5) reviewing
27   environmental data, and (6) observing environmental sampling.
28
29   Under DOE Order 435.1, DOE must maintain its performance assessment, which is used as the
30   primary basis for demonstrating that the performance objectives will be met during the
31   compliance period. Performance assessment maintenance can include conducting research,
32   implementing field studies, and performing monitoring needed to address uncertainties or gaps
33   in existing data. Under DOE Order 435.1, DOE must annually determine the continued
34   adequacy of its performance assessment based upon the results of its data collection and
35   analysis from its research, field studies, and monitoring. In addition, DOE must provide annual
36   summaries of conclusions and recommendations from its performance assessment and identify
37   any needed revisions. Because some of the data collected and/or research undertaken by DOE
38   to maintain its performance assessment could provide additional support for some of the key
39   aspects identified by the staff in its TER, DOE's activities are expected to have a direct bearing
40   on NRC's monitoring activities. In addition, under DOE Order 435.1, DOE must develop a
41   monitoring plan for internal use, and NRC staff may request a copy of this plan. The
42   environmental monitoring plan is designed to address measurement and evaluation of releases,
43   migration of radionuclides, disposal unit subsidence, and changes in disposal facility and
44   disposal site parameters that may affect long-term performance.
45
46   A key part of NRC's monitoring plan development is coordinating with the covered State to
47   identify monitoring activities that will help NRC perform its responsibilities under the NDAA. To
48   the extent that it is able, NRC will take full advantage of the covered State's regulatory programs
49   related to monitoring of the disposal facility. For example, the covered State may have specific

1 requirements related to well construction and sampling that may help NRC ensure that wells are
2 properly installed and reliable samples are collected and analyzed.
3
4 Once NRC develops a compliance monitoring plan, a copy will be provided to DOE and the
5 covered State. Further, compliance monitoring plans will be publicly available and accessible
6 through NRC's website. After NRC has developed its monitoring plan for a specific waste
7 determination, it will perform the activities described in the plan. Plan implementation may
8 begin with NRC sending a written request to DOE for the data, studies, or analyses identified in
9 the monitoring plan. NRC staff will also request that DOE provide a schedule of its waste
10 disposal activities. After NRC has received the requested information from DOE, staff will
11 technically review the information. In addition, staff should coordinate an onsite observation
12 schedule with DOE and the covered State.
13
14 If NRC is unable to obtain needed information from DOE in order to complete NRC's monitoring
15 activities, key assumptions and aspects relied upon to demonstrate compliance with the
16 performance objectives may no longer be adequately supported, and NRC staff may no longer
17 have reasonable assurance that the performance objectives will be met. In this case, NRC will
18 inform DOE, Congress and the covered State by issuing a Type III noncompliance notification
19 letter (see Section 10.4.3).
20
21 If, during the technical reviews or onsite observations, NRC staff obtains data and information
22 that indicate that key assumptions and aspects relied upon to demonstrate compliance with one
23 or more performance objectives are no longer adequately supported, the staff will evaluate the
24 information to determine its significance in the context of potential noncompliance with a
25 performance objective. Staff will need to evaluate whether one or more of the performance
26 objectives will likely not be met. Even if a key assumption relied upon to demonstrate
27 compliance with one or more performance objectives is no longer supported, DOE may be able
28 to demonstrate that the performance objective(s) will be met by demonstrating that the key
29 assumption is compensated by conservative assumptions in other parts of the analysis. It is
30 anticipated that the performance assessment overall may be an iterative process because an
31 individual component's importance can be relative to the performance of other components
32 (e.g., subsequent iterative analyses would need to clearly show that an unsupported
33 assumption would cause the performance objective[s] not to be met). Understanding the
34 importance of parameters and model components and their interrelationships with one another
35 is best accomplished by exploring a variety of approaches. Accordingly, the staff will need to
36 confer with DOE before concluding that there is no longer reasonable assurance that a
37 performance objective will be met in the future. Once the staff determines that there is no
38 longer reasonable assurance that a performance objective will be met, NRC has an obligation to
39 notify Congress, the covered State, and DOE as soon as practicable in the form of a notification
40 letter (Type II) from the Chairman of NRC. A Type I letter would state that there are sufficient
41 indications that DOE is currently not meeting the requirements of 10 CFR 61.41–61.44. The
42 types of notification letters are discussed in more detail in Section 10.4.3.
43
44 NRC staff should document the findings of its assessment in a periodic compliance monitoring
45 report. At this time the staff intends to develop these periodic reports on an annual basis (see
46 Section 10.4.2). This periodic report should identify the status of monitoring activities
47 (i.e., open, closed, open-noncompliant). In addition, the periodic report will identify changes
48 that should be made to NRC's monitoring activities for the upcoming year. NRC will also
49 request that DOE provide an updated schedule of waste disposal activities for NRC planning
50 purposes (e.g., to enable the observation of disposal actions). The periodic compliance

1 monitoring report should document any significant changes to the disposal design. If the
2 changes are significant, the periodic monitoring report should recommend that the compliance
3 monitoring plan be revised. It is probable that the compliance monitoring plan will need to be
4 updated and revised as plans for the disposal site are revised or finalized. Monitoring activities
5 should focus on risk-significant features and processes that are part of finalized designs for
6 engineered barriers and not on the details for designs that are still tentative or under
7 development. For example, a final decision on erosion and infiltration controls may not be made
8 until years after disposal actions involving grouting are expected to be completed. Similarly,
9 new designs not considered in the original waste determination may be developed based on
10 new research. Updating monitoring activity details during revisions to the monitoring plan
11 would, therefore, need to reflect and emphasize the revised or final design plans.
12
13 The NDAA does not specify how long the NRC is responsible for assessing compliance,
14 therefore, the monitoring activity is indefinite. In general, it is envisioned that the level of
15 NRC-required effort will vary over time, with an initially higher level of effort needed early in the
16 process to review information and observe actions that may potentially support key assumptions
17 and parameters made as part of DOE's disposal actions. As the monitoring activities close, the
18 scope of NRC's monitoring efforts should be reduced. Figure 10-1 shows the general process
19 the Commission will follow to carry out its monitoring responsibilities and when issuing Type II
20 and III noncompliance notification letters.
21

22 **10.2   Coordinating With the Covered State**
23
24 As previously stated, a key part of NRC's monitoring responsibilities under the NDAA is to
25 coordinate monitoring activities with the covered State. As part of the general monitoring
26 process, it is recommended that discussions with the covered State concerning monitoring
27 occur during the waste determination review. Discussions with the State will enable NRC staff
28 to determine the type of activities the State will be undertaking in terms of monitoring
29 compliance with its regulatory requirements and enable the covered State to understand the
30 type of activities that NRC will be undertaking. Insights gained through these discussions may
31 help the two parties coordinate their activities in order for NRC to meet its NDAA monitoring
32 responsibilities. In some cases, NRC may be able to rely upon State-obtained information.
33 NRC will need to periodically communicate with the State to ascertain any changes In the
34 State's permitting requirements because changes may affect the type and frequency of some of
35 the DOE-collected data.
36
37 NRC staff will keep the covered State informed of its monitoring activities. NRC staff should, at
38 appropriate times, observe specific DOE activities at the site and review records. The covered
39 State will be notified of planned onsite observations and may participate if it so decides. Onsite
40 observation reports summarizing findings and results will be provided to the covered State.
41
42 NRC will provide a draft copy of the monitoring plans for the State's comment prior to finalizing
43 the plans. NRC will also share with the State any letters sent to DOE indicating possible
44 compliance concerns. The staff expects to meet periodically with DOE and the covered State
45 throughout the early stages of monitoring to ascertain the need for changes to the NRC's
46 compliance monitoring plan. These meetings will provide opportunities to discuss the status of
47 NRC monitoring activities and identify potential problems. Further, these meetings will provide
48 opportunities to discuss specific monitoring activities (e.g., onsite observations) planned for the
49

1
2

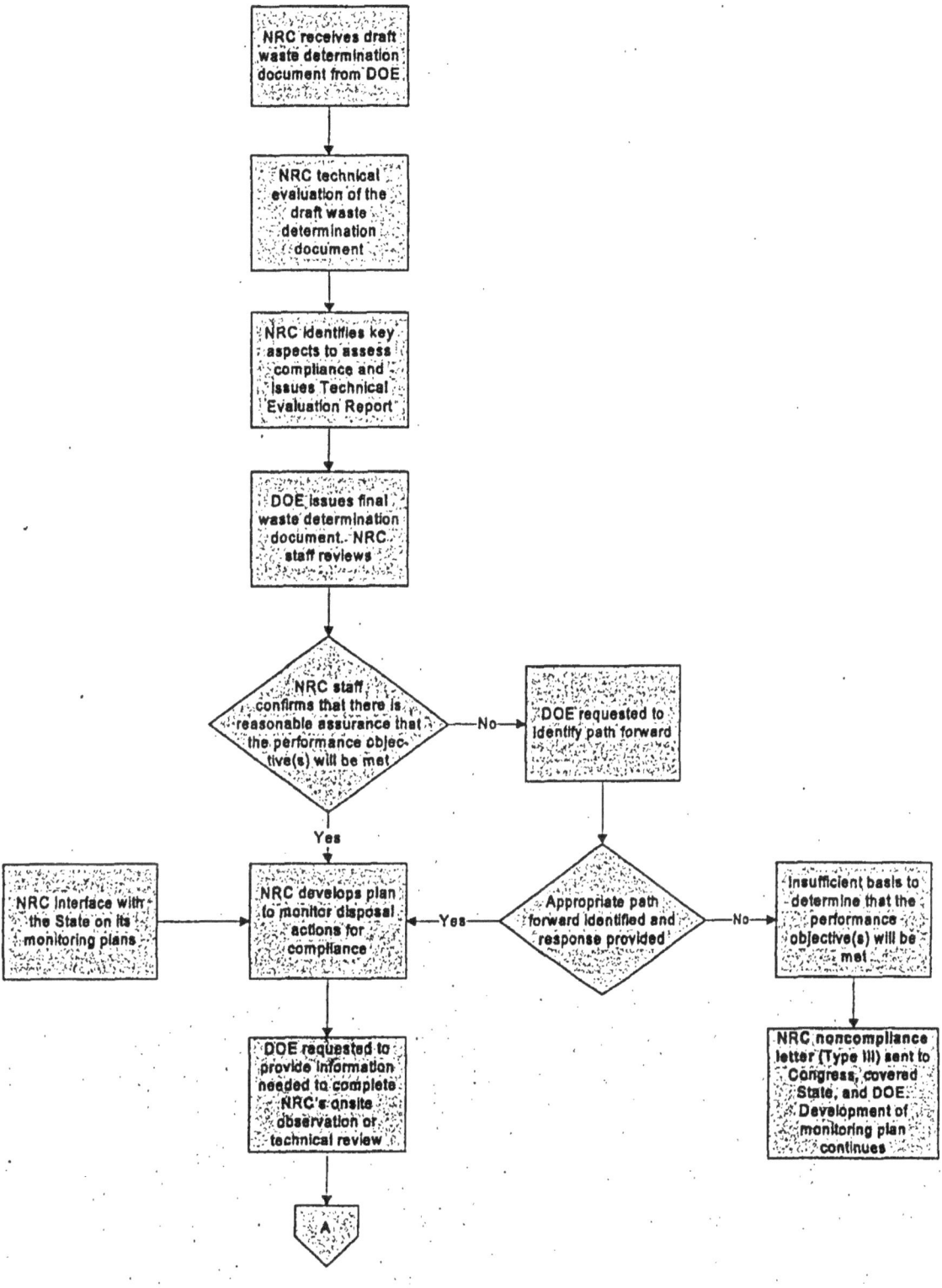

3 **Figure 10-1. Schematic of NRC's Compliance Monitoring Approach**
4

1
2

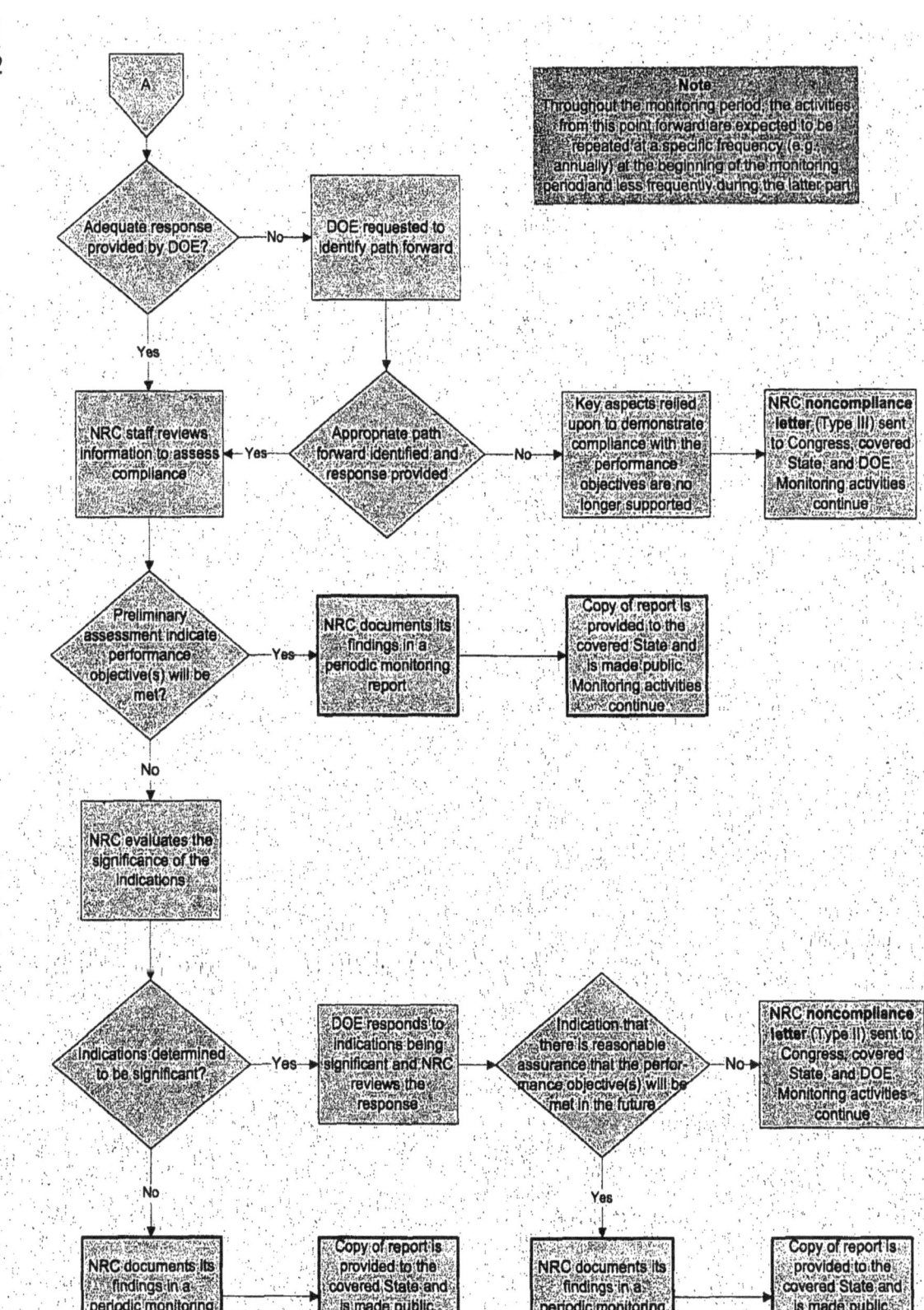

Figure 10-1 (continued). Schematic of NRC's Compliance Monitoring Approach

3

1 upcoming year, and allow the State to identify how it may want to participate in the
2 planned activities.
3
4 ## 10.3   Compliance Monitoring Activities
5
6 The objective of NRC's monitoring is to assess whether DOE's disposal actions are in
7 compliance with the performance objectives in 10 CFR Part 61, Subpart C. Subpart C contains
8 a general requirement (10 CFR 61.40), requirements for protection of the general population
9 from releases of radioactivity (10 CFR 61.41), protection of individuals from inadvertent intrusion
10 (10 CFR 61.42), protection of individuals during operations (10 CFR 61.43), and stability of the
11 disposal site after closure (10 CFR 61.44). NRC staff will focus monitoring activities on those
12 waste disposal aspects that may affect whether the performance objectives can be met at the
13 present time and in the future.
14
15 The status of technical review and onsite observation activities is expected to fall into one of
16 three categories: (1) closed, (2) open, or (3) open-noncompliant. Staff will only monitor
17 activities that are categorized as open or open-noncompliant. The distinction between the last
18 two categories is made primarily to distinguish between monitoring activities that are merely
19 ongoing and monitoring activities that are ongoing and about which NRC has issued a
20 notification letter of noncompliance. A simplified example follows: if NRC staff reviewed data
21 used to support DOE's assumption(s) regarding the creation of reducing conditions in the
22 wasteform (due to the addition of ground blast furnace slag to the grout), and NRC staff
23 concluded that DOE's assumption could no longer be supported, then, after a noncompliance
24 notification letter had been sent, this monitoring activity would be given the status of
25 open-noncompliant pending further review and discussion with DOE. Alternatively, DOE could
26 demonstrate reducing conditions in the grouted wasteform are no longer necessary for
27 demonstrating compliance with the performance objectives such that NRC could conclude with
28 reasonable assurance that the disposal facility will meet the performance objective(s) and that
29 the monitoring activity status is open while disposal action is ongoing.
30
31 In the periodic compliance monitoring report, the technical review activities and the onsite
32 observation activities will be given a status of either closed, open, or open-noncompliant.
33 This is the tracking mechanism that allows NRC staff to quickly see which activities have been
34 concluded and which activities need special attention. Each monitoring activity, whether
35 technical review or onsite observation, is associated with a disposal action and should be so
36 identified in the relevant compliance monitoring plan. The active phase of the disposal actions
37 occurs during the time of operations that NRC defines as the time during which DOE carries out
38 its waste disposal actions, through the end of the institutional control period. Waste disposal
39 actions are considered to include performance assessment revisions, waste removal, grouting,
40 stabilization, observation, maintenance, or other similar activities. A technical review activity
41 that requires NRC staff to review information on curing techniques for grout would be associated
42 with the disposal action of grouting, while an onsite observation activity that requires NRC staff
43 to observe the construction of an engineered surface cover would be associated with the
44 disposal action of stabilization. For example, assume the results of the technical review for
45 curing techniques could no longer support key aspects of grouting, and there was no
46 reasonable assurance that one or more of the performance objectives would be met. NRC
47 would send a noncompliance notification letter stating that the grouting disposal action is not in
48 compliance with the performance objectives of 10 CFR Part 61 even if the other monitoring
49 activities relating to grouting have a status other than noncompliant.

1   The NRC staff should accurately track the status of open or open-noncompliant activities in the
2   periodic compliance monitoring report (see Section 10.4.2). As DOE's performance
3   assessment is revised and updated, it is conceivable that new issues that NRC staff determines
4   are important to DOE's compliance demonstration will be identified and should be monitored.
5   Therefore, as NRC proceeds with its monitoring activities, it is anticipated that new monitoring
6   activities could be developed and that the list of monitoring activities could change as some
7   activities are closed and new activities are opened. A closed activity can be reopened if new
8   information becomes available.
9
10 If monitoring activities provide information that there is no longer reasonable assurance that the
11 performance objectives will be met, then that monitoring activity may be close to being
12 categorized as open-noncompliant. If NRC staff determines that DOE is not in compliance with
13 one or more performance objective(s), staff should document its findings so that this can be
14 conveyed to DOE. DOE should be afforded an opportunity to provide additional information,
15 analyses, and/or insights that could help the staff reach a final conclusion. If it is still
16 determined that there is no reasonable assurance that one or more performance objectives will
17 be met, NRC staff will categorize the monitoring activity as open-noncompliant and send out a
18 notification letter that DOE is not in compliance (see Section 10.4.3).
19

20 **10.3.1   Technical Review Approach**
21

22 This section of the guidance document describes the general approach that the staff will take in
23 its review of technical information such as data, studies, experiments, and analyses specifically
24 identified to be addressed during monitoring. NRC technical review activities will consist of two
25 primary activities: reviewing data related to DOE's waste disposal actions and reviewing DOE's
26 performance assessment. Along with reviewing DOE's performance assessment, NRC staff will
27 review studies and analyses that support the performance assessment. The level of detail of
28 the performance assessment review will depend on whether DOE revises its performance
29 assessment, the extent of these changes, and how these changes and their effects are
30 documented and referenced. Most of NRC monitoring activities related to technical reviews (as
31 opposed to onsite observations) will be designed to assist in assessing compliance with
32 10 CFR 61.41 and 61.42 because the compliance demonstration relies upon long-term
33 predictive analyses. Most NRC technical reviews will focus on addressing uncertainties in these
34 predictive analyses. In review of data, staff will need to ensure that there is reasonable
35 confidence in the quality of the data in terms of its traceability, reproducibility, and
36 representativeness. Thus, the staff may need to review the approach DOE used to ensure the
37 quality of its data.
38

39 Environmental data should be sampled frequently and periodically and subsequently evaluated
40 and compared to an established metric that can be used to identify potential concerns. The
41 charting approach recommended by the U.S. Environmental Protection Agency (EPA) for
42 determining the adequacy of cleanup at Superfund sites may be useful for deciding when the
43 environmental data indicates a potential problem (EPA, 1989). Under the EPA charting
44 approach, the data for each sampling location would be charted as a function of time. A
45 potential concern would be indicated when either a specific number of consecutive samples
46 exceed the long-term average (or predisposal average) or there are a specific number of
47 consecutive increases in samples. This approach, used in the way control charts detect
48 significant deviations in an industrial process, is easy to use and should provide a fairly reliable
49 means of discerning an early indication of potential problems. In applying control charts for
50 industrial processes, eight consecutive numbers are generally used to indicate a change in the

1 process. This results in a very low probability of incorrectly concluding that there is a change in
2 the process (0.8 percent), assuming the data are independent. For environmental data, this
3 level of precision is not needed; therefore, it is generally recommended to use five to seven
4 consecutive points to indicate of a change in the system (EPA, 1989). Because the intent is
5 only to provide an early indication of potential concerns, an even lower precision is warranted;
6 therefore, it is recommended that the staff use 3 to 4 consecutive points to indicate potential
7 concerns. Figure 10-2 provides an example of a control chart. In this example, four
8 consecutive data points are above the predisposal average and thus would be interpreted as an
9 early indication of a potential problem. Note that in the example, the contaminant concentration
10 levels are below the concentration that would give a dose of 25 mrem/year. However, because
11 four consecutive points are above the long-term average, this should be interpreted as a
12 change in the system that warrants further evaluation.
13
14 Key aspects identified within the TER for the waste determination will need to be assessed
15 through the review of data, studies, experiments, and/or analyses carried out by DOE. Staff
16 should review available studies and reports to see whether conclusions DOE reached continue
17 to be supported. In reviewing results from experiments, the staff should consider whether or not
18 appropriate approaches were used, and the results can be extrapolated to a larger scale and/or
19 over a longer time period. In reviewing DOE analyses, especially DOE's performance
20 assessment, the staff should ensure that sufficient data continues to support assumptions within
21 the analyses and that the appropriate range of data to account for variability, the appropriate
22 consideration of model uncertainty, and the adequacy of the model support is still valid.
23 Modeling results should have either adequate model support or appropriate conservative
24 assumptions or parameter values. In general, the results from one model should not be used as
25 the sole support for another model. The amount of support that is provided for models that form
26 the basis for conclusions is a key issue. In reviewing DOE analyses, the staff should consider
27
28
29

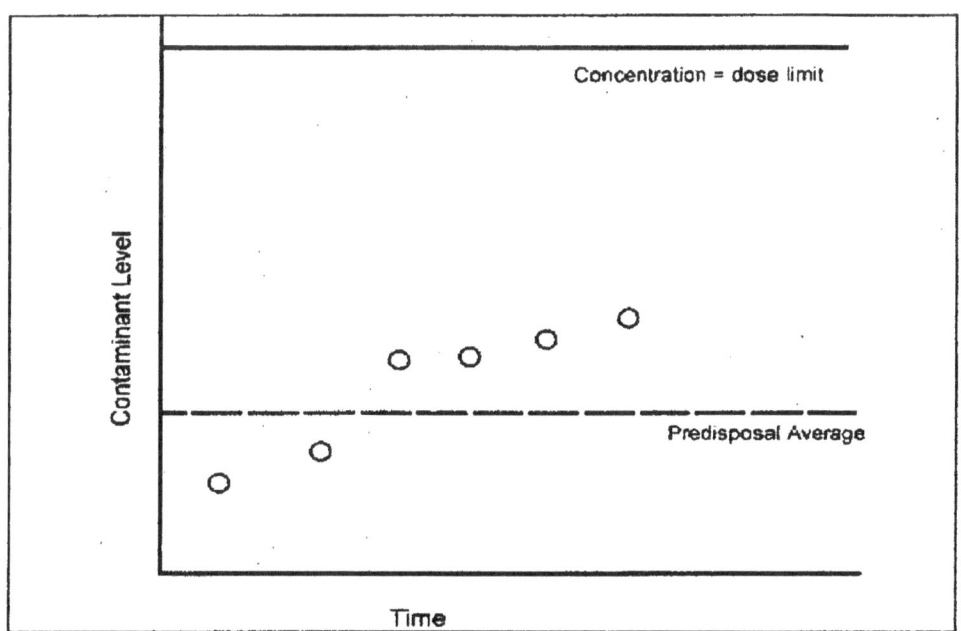

30   **Figure 10-2. A Control Chart Example**

1   individual analyses in the context of the overall performance demonstration (i.e., how well the
2   analyses comport with other analyses and the significance of the results from an individual
3   analysis in the context of the overall performance demonstration). To the extent practical, staff
4   should perform its own independent analyses to substantiate results from DOE's analysis and to
5   understand the important aspects of the analysis.
6
7   As staff completes its technical review and depending on the type of monitoring activity,
8   compliance monitoring activities may be discontinued and given a status of closed following the
9   approach described in the specific compliance monitoring plans. Any activity that is closed will
10 be documented as such in the periodic compliance monitoring report (see Section 10.4.2).
11

12 ## 10.3.2   Onsite Observation Approach

13
14 As previously mentioned, there are generally two primary approaches staff uses to assess
15 compliance with the performance objectives in 10 CFR Part 61, Subpart C: technical reviews
16 and onsite visits or observations. In coordination with the covered State, NRC staff will, at
17 appropriate times, go to the site to observe and review certain operations as they are being
18 performed. NRC onsite observations may include activities such as (1) observation and review
19 of data collection activities, (2) observation of grout preparation and pouring, (3) review of
20 radiation records and associated program elements, and (4) observation of construction of
21 engineered features and their maintenance. Onsite observations should be conducted using a
22 performance-based approach. Observations should focus on direct observation of work
23 activities, interviews with workers, demonstrations of tasks, visual review of the construction of
24 facilities, and where appropriate, a review of selected records. A direct examination of these
25 activities and discussions with cognizant workers will help provide important insights and is
26 preferable to a review of selected records alone. NRC staff must be prepared to meet all
27 DOE-established entry requirements. This includes, but may not be limited to, use of personal
28 protective equipment, viewing of a site-specific safety video, or meeting any special
29 requirements for entering certain environments.
30
31 The frequency of periodic observations of DOE's waste disposal activities may be dependent on
32 DOE plans and should be selected based on the stage of waste disposal. DOE's radiation
33 protection program is expected to be sufficiently mature and is expected to have had
34 demonstrated adequate protection. Therefore, these observations involving protection of
35 individuals during operations will focus primarily on confirming that DOE radiation protection
36 plans for the specific waste disposal activities are consistent with the overall radiation protection
37 program cited in the waste determination as the basis for demonstrating compliance with the
38 10 CFR 61.43 performance objective. After the NRC staff has gained an understanding of the
39 scope of the radiation protection program, future assessments may be greatly scaled back and
40 focus primarily on program changes.
41

42 ## 10.4   Documentation of Findings and Reporting

43
44 ## 10.4.1   Onsite Observation Reports

45
46 NRC staff participating in onsite observations will be required to develop a report after each site
47 visit. This report will provide a written summary of what was done on the site visit and list any
48 results obtained through the activities performed at the site. The report is expected to include a
49 list of the participants, a description of the activities undertaken, and staff assessment of

1 materials evaluated or direct observations. This report will also describe which onsite
2 observation activities identified in the compliance monitoring plan were carried out during the
3 onsite observation. In addition, the report should clearly indicate which activities remain open
4 and are to be continued in the future and which have been satisfactorily completed and given
5 the status of closed. The report of the onsite monitoring results will be provided to DOE and
6 covered State by letter within 60 days following the site visit.
7
8 ## 10.4.2   Periodic Compliance Monitoring Reports
9
10 Staff will periodically document its findings for the various technical reviews and onsite
11 observations completed in a monitoring report.[1] In addition to the onsite observation findings
12 made during the year, the periodic report will also document the staff's technical reviews of
13 data, its review of reports or analyses to address key aspects, and its review of any updates or
14 revisions to DOE's performance assessment.
15
16 In documenting its review, the staff will describe what was reviewed, its findings on whether
17 there are indications that there is no longer reasonable assurance that the performance
18 objectives will be met for the entire performance period (this will include a description of
19 the specific performance objectives that apply), and the basis for the staff findings
20 (e.g., independent analyses conducted by the staff, supporting studies, expert opinion). In
21 addition, the staff should describe any recommended action that should be undertaken to close
22 open activities.
23
24 The periodic compliance monitoring report should also serve as a vehicle for tracking the status
25 of the monitoring activities in terms of whether the activity is considered closed, open, or
26 open-noncompliant. For any activity that was previously closed but is being reopened, the staff
27 should describe its basis for reopening the activity and its expected plans for monitoring the
28 activity. The periodic compliance monitoring report should also describe any new monitoring
29 activities identified during the year and the basis for opening the new activity. For open and
30 open-noncompliant activities, the staff should also identify any activities that are expected to be
31 carried out for the upcoming year. For activities that have been closed, the staff should
32 document its basis for closing the review and any conditions that could prompt its reopening.
33 For noncompliant activities, the periodic compliance monitoring report should document any
34 actions or results that might change the status. In addition to giving the status of each
35 monitoring activity, the periodic compliance monitoring report should identify which disposal
36 action each monitoring activity is associated with, as previously documented in the compliance
37 monitoring plan (see Section 10.3).
38
39 It is envisioned that staff will meet with DOE and the State on an periodic[2] basis to discuss the
40 status of the monitoring program. The staff may use this meeting to help identify potential
41 revisions to the monitoring program.[3] NRC will take the lead in setting up such a meeting,

---

[1]It is anticipated that during the early phases of carrying out its monitoring activities, the NRC will develop an annual report. However, as the monitoring program progresses and the number of monitoring activities diminish, the staff will need to reassess whether less frequent reporting is warranted.

[2]The frequency of these meetings is expected to change as implementation of the NRC monitoring program progresses. As the number of monitoring activities diminish, NRC staff will reevaluate the meeting frequency.

[3]This could also include determining the need for less frequent reporting and meetings.

1 including developing the agenda and making the meeting notices (these meetings are expected
2 to be open to the public). Given that there may be some public interest in such meetings, some
3 of the meetings may be held in the vicinity of the site; therefore, the NRC staff may need to work
4 with DOE and/or the State in securing an appropriate meeting facility. Staff should describe
5 conclusions and followup actions that come out of this meeting in the periodic compliance
6 monitoring report.
7
8 The monitoring report should describe any developing issues or potential problems that should
9 be brought to the attention of management. For example, the staff may have identified issues
10 through either its technical review or onsite observations that are not significant enough to
11 indicate that there is no longer reasonable assurance that the performance objectives will be
12 met, but issues that could potentially affect the performance demonstration may be present.
13 Such issues will be discussed through normal correspondence for issues associated with
14 technical reviews, and through exit meetings and onsite observation reports for issues identified
15 during onsite observations. In addition, developing issues or potential problems identified
16 through technical reviews or onsite observations will be included in the annual monitoring
17 reports. The disposition of any issues raised during the previous year should also be described
18 in the report.
19
20 Figure 10-3 shows the topical areas that are likely to be covered in the periodic compliance
21 monitoring report. A copy of the report will be provided to DOE and the State for information
22 purposes. In addition, the report will be made publicly available on NRC's website.
23
24 **10.4.3   Noncompliance Notification Letters**
25
26 In accordance with the NDAA, NRC is required to inform DOE, the covered State, and Congress
27 if it considers any of DOE's waste disposal actions to be noncompliant with the performance
28 objectives of 10 CFR Part 61, Subpart C. The specific congressional committees that NRC is
29 required to inform are the Committee on Armed Services, Energy and Commerce, and
30 Appropriations in the House of Representatives, and the Committee on Armed Services, Energy
31 and Natural Resources, Environment and Public Works, and Appropriations in the Senate. In
32 addition, the noncompliance notification letter will be made publicly available on NRC's website.
33 NRC is required to make this notification as soon as practicable after discovery of
34 noncompliant conditions.
35
36 As NRC staff carries out its monitoring activities as outlined in the compliance monitoring plan,
37 preliminary assessments will indicate whether DOE's disposal actions are considered to be in
38 compliance with the performance objectives. There are three primary reasons that DOE
39 disposal actions could be found noncompliant: (1) if there are sufficient indications that the
40 requirements of 10 CFR 61.41 through 44 are currently not being met, (2) if there are sufficient
41 indications that there is no longer reasonable assurance that the performance objectives will be
42 met in the future, or (3) if key aspects relied upon to demonstrate compliance with one or more
43 performance objectives are no longer supported due to the lack of supporting information
44 obtained during the monitoring period.
45
46 With respect to the first primary reason, possible indications that the requirements of
47 10 CFR 61.41 through 44 are currently not being met would be environmental concentrations at
48 locations where individuals could be exposed that would give a dose exceeding the dose criteria
49 set by the performance objective(s). Other possible indications that the performance objectives

Topical Areas

Onsite Observations
- Areas reviewed
- Preliminary assessments
- Basis for preliminary assessments
- Recommended actions

Technical Reviews
- Areas reviewed
- Preliminary assessments
- Basis for preliminary assessments
- Recommended actions

New and Reopened Activities
- Area of concern
- Significance to performance demonstration
- Expected monitoring activities

Open-Noncompliant Activities
- Basis for status
- Actions or results that might change the status

Table Matching Each Monitoring Activity with Disposal Action

Summary of Periodic Meeting

Revisions to the Disposal System Designs and the Compliance Monitoring Plans

Potential Problems

Activities Closed

**Figure 10-3. Topical Areas Expected in the Periodic Compliance
Monitoring Report**

1 are currently being exceeded would be radiation doses to workers or members of the public that
2 exceed the dose limit.

3 With respect to the second reason, NRC staff note that 10 CFR 61.40 is not met as soon as
4 there is no longer reasonable assurance that the requirements of 10 CFR 61.41–61.44 will not
5 be met at any time during the performance period. Given the nature of the highly engineered
6 facilities involved, evidence of problems with meeting the performance objectives may not be
7 observable for hundreds of years in the future. Thus, solely relying upon observable failure of
8 the system may not allow the NRC to make a timely notification as required by the NDAA. In
9 addition, assessing compliance for some performance objectives (e.g., 10 CFR 61.42) is difficult
10 to accomplish through direct observation. Thus, the second means by which NRC may make a

1　finding of noncompliance is through predictive modeling that indicates that there is no longer
2　reasonable assurance that the performance objectives will be met for the entire performance
3　period. For example, predicted or expected technical information on a key assumption was
4　replaced with conflicting new information. Key in this sense means that without the assumption,
5　demonstrating compliance with the performance objectives may not be possible. Another
6　indication would be if trends in the data indicate that at some future time the performance
7　objectives criteria would be exceeded.
8
9　The third way in which the performance objectives may not be met is when key aspects relied
10　upon to demonstrate compliance with one or more performance objectives are no longer
11　supported due to the lack of supporting information obtained during the monitoring period (see
12　Section 10.1).
13
14　Given the different types of noncompliance, NRC anticipates using several different types of
15　notification letters. These letters are listed in Table 10-2. A Type I letter would state that there
16　are sufficient indications that the requirements of 10 CFR 61.41 through 44 are currently not
17　being met. Within the letter, NRC would document the basis for the indications that the
18　performance objectives are currently not being met and describe the performance objectives. A
19　Type II letter would document the basis for the indication that there is no longer reasonable
20　assurance that the performance objectives will be met for the entire performance period and
21

Table 10-2. Types of Notification Letters			
Type	**Notification**	**Signature**	**Distribution**
I	Noncompliance Indication that the requirements of 10 CFR 61.41 through 44 are *currently* not being met	Chairman	DOE, covered State, and Congress
II	Noncompliance Indication that there is no longer reasonable assurance that the performance objective(s) will be met in the *future*	Chairman	DOE, covered State, and Congress
III	Noncompliance *Lack of supporting information*: Key aspects relied upon to demonstrate compliance with the performance objective(s) are no longer supported	Chairman	DOE, covered State, and Congress
IV	Concern Concerns with the performance demonstration	NRC Staff Management	DOE and covered State
V	Resolution Resolution of concerns with the performance demonstration	NRC Staff Management	DOE and covered State

29

1   would describe the specific performance objectives that may not be met. A Type III letter would
2   document the lack of supporting information during the compliance monitoring period. This
3   letter would include the basis for such a conclusion and describe the performance objective(s)
4   lacking supporting information. Because of their significance and distribution, Type I–III letters
5   would be sent out under the signature of the NRC Chairman. While each of the three types of
6   notification letters is important, the Type I letter is the most serious because it pertains to an
7   immediate potential threat to public health and safety.
8
9   Prior to sending out Type I–III letters, NRC will review its concerns in a letter (Type IV) to DOE
10   and the State. This will give the State an opportunity to provide input and comment and provide
11   DOE with an opportunity to provide information that demonstrates its disposal actions are in
12   compliance with the performance objectives. Assuming that DOE provides information to
13   support its performance demonstration, NRC will review this information and decide whether it is
14   sufficient to conclude that there is reasonable assurance that the performance objectives will be
15   met. If the staff determines that, based on the information DOE provided, there is sufficient
16   basis to conclude that DOE is in compliance, NRC will send out a notification of resolution letter
17   (Type V). Type IV and V letters will be made publicly available on NRC's website. These letters
18   formally document resolution of the issue. If the staff determines that, based on the information
19   DOE provides, there is still a basis for concluding that DOE is noncompliant, NRC will send out
20   the noncompliance notification letter (i.e., Type I–III).
21
22   Because NRC does not have regulatory or enforcement authority over DOE, it is the role of
23   Congress, the covered State, and DOE to determine what, if any, actions will be taken in
24   response to a noncompliance notification letter. The noncompliance notification letter will
25   emphasize that the status of the disposal action may change to something other than
26   noncompliant depending on the type of information NRC receives in the future. Any subsequent
27   change in the status of the disposal action from noncompliant will be documented in the periodic
28   compliance monitoring report, which will be publicly available. NRC staff will continue to monitor
29   DOE's disposal actions after issuance of the noncompliance notification letter.
30

31   **10.5   References**

32
33   U.S. Environmental Protection Agency (EPA). "Statistical Analysis of Ground-Water Monitoring
34   Data at RCRA Facilities, Interim Final Guidance." Office of Solid Waste, U.S. Environmental
35   Protection Agency. Washington, DC. February 1989.
36
37   ———. "Decommissioning Criteria for the West Valley Demonstration Project at the West Valley
38   Site: Final Policy Statement." *Federal Register*, 67 FR 5003. February 2002.
39
40   U.S. Nuclear Regulatory Commission (NRC). "Implementation of New U.S. Regulatory
41   Commission Responsibility under the National Defense Authorization Act of 2005 in Reviewing
42   Waste Determinations for the U.S. Department of Energy." SECY–05–0073. April 28, 2005.
43
44
45

APPENDIX A
SECTION 3116
OF THE RONALD W. REAGAN NATIONAL DEFENSE AUTHORIZATION ACT FOR FISCAL YEAR 2005

SEC. 3116. DEFENSE SITE ACCELERATION COMPLETION.

(a) IN GENERAL- Notwithstanding the provisions of the Nuclear Waste Policy Act of 1982, the requirements of section 202 of the Energy Reorganization Act of 1974, and other laws that define classes of radioactive waste, with respect to material stored at a Department of Energy site at which activities are regulated by a covered State pursuant to approved closure plans or permits issued by the State, the term "high-level radioactive waste" does not include radioactive waste resulting from the reprocessing of spent nuclear fuel that the Secretary of Energy (in this section referred to as the "Secretary"), in consultation with the Nuclear Regulatory Commission (in this section referred to as the "Commission"), determines—

    (1) does not require permanent isolation in a deep geologic repository for spent fuel or high-level radioactive waste;

    (2) has had highly radioactive radionuclides removed to the maximum extent practical; and

    (3)(A) does not exceed concentration limits for Class C low-level waste as set out in section 61.55 of title 10, Code of Federal Regulations, and will be disposed of—

        (i) in compliance with the performance objectives set out in subpart C of part 61 of title 10, Code of Federal Regulations; and

        (ii) pursuant to a State-approved closure plan or State-issued permit, authority for the approval or issuance of which is conferred on the State outside of this section; or

    (B) exceeds concentration limits for Class C low-level waste as set out in section 61.55 of title 10, Code of Federal Regulations, but will be disposed of—

        (i) in compliance with the performance objectives set out in subpart C of part 61 of title 10, Code of Federal Regulations;

        (ii) pursuant to a State-approved closure plan or State-issued permit, authority for the approval or issuance of which is conferred on the State outside of this section; and

        (iii) pursuant to plans developed by the Secretary in consultation with the Commission.

(b) MONITORING BY NUCLEAR REGULATORY COMMISSION- (1) The Commission shall, in coordination with the covered State, monitor disposal actions taken by the Department of Energy pursuant to subparagraphs (A) and (B) of subsection (a)(3) for the purpose of assessing compliance with the performance objectives set out in subpart C of part 61 of title 10, Code of Federal Regulations.

(2) If the Commission considers any disposal actions taken by the Department of Energy pursuant to those subparagraphs to be not in compliance with those performance objectives, the Commission shall, as soon as practicable after discovery of the noncompliant

conditions, inform the Department of Energy, the covered State, and the following congressional committees:

(A) The Committee on Armed Services, the Committee on Energy and Commerce, and the Committee on Appropriations of the House of Representatives.

(B) The Committee on Armed Services, the Committee on Energy and Natural Resources, the Committee on Environment and Public Works, and the Committee on Appropriations of the Senate.

(3) For fiscal year 2005, the Secretary shall, from amounts available for defense site acceleration completion, reimburse the Commission for all expenses, including salaries, that the Commission incurs as a result of performance under subsection (a) and this subsection for fiscal year 2005. The Department of Energy and the Commission may enter into an interagency agreement that specifies the method of reimbursement. Amounts received by the Commission for performance under subsection (a) and this subsection may be retained and used for salaries and expenses associated with those activities, notwithstanding section 3302 of title 31, United States Code, and shall remain available until expended.

(4) For fiscal years after 2005, the Commission shall include in the budget justification materials submitted to Congress in support of the Commission budget for that fiscal year (as submitted with the budget of the President under section 1105(a) of title 31, United States Code) the amounts required, not offset by revenues, for performance under subsection (a) and this subsection.

(c) INAPPLICABILITY TO CERTAIN MATERIALS- Subsection (a) shall not apply to any material otherwise covered by that subsection that is transported from the covered State.

(d) COVERED STATES- For purposes of this section, the following States are covered States:

(1) The State of South Carolina.

(2) The State of Idaho.

(e) CONSTRUCTION- (1) Nothing in this section shall impair, alter, or modify the full implementation of any Federal Facility Agreement and Consent Order or other applicable consent decree for a Department of Energy site.

(2) Nothing in this section establishes any precedent or is binding on the State of Washington, the State of Oregon, or any other State not covered by subsection (d) for the management, storage, treatment, and disposition of radioactive and hazardous materials.

(3) Nothing in this section amends the definition of "transuranic waste" or regulations for repository disposal of transuranic waste pursuant to the Waste Isolation Pilot Plant Land Withdrawal Act or part 191 of title 40, Code of Federal Regulations.

(4) Nothing in this section shall be construed to affect in any way the obligations of the Department of Energy to comply with section 4306A of the Atomic Energy Defense Act (50 U.S.C. 2567).

(5) Nothing in this section amends the West Valley Demonstration Act (42 U.S.C. 2121a note).

(f) JUDICIAL REVIEW- Judicial review shall be available in accordance with chapter 7 of title 5, United States Code, for the following:

(1) Any determination made by the Secretary or any other agency action taken by the Secretary pursuant to this section.

(2) Any failure of the Commission to carry out its responsibilities under subsection (b).

APPENDIX B
CALCULATION TO DEVELOP BENCHMARK AVERAGING EXPRESSIONS

This section describes the calculations used to develop the benchmark expressions for Category 3 concentration averaging. The primary purpose of the example averaging expressions is to provide a benchmark for the staff to use when reviewing site-specific averaging for waste classification. The benchmark provides a comparison point for what can be complex calculations involving numerous uncertain parameters and assumptions. This appendix is not intended to provide a full description of an internal staff review tool. A full description may be developed at a later date if warranted and if sufficient resources are available.

B.1 Development of Example Averaging Expressions

A description of the calculations used to develop the example averaging expressions for Category 3 are provided in this section. Conceptually, the approach is based on the following equation

$$C_{i,j} * V_i * X_{i,j} = D_{i,j} \qquad \text{(B–1)}$$

where

i	=	the analysis index (set to either 61 or N)
j	=	the radionuclide index
$C_{i,j}$	=	the concentration of radionuclide j for analysis i
V_i	=	the volume of waste exhumed for analysis i
$X_{i,j}$	=	a conversion factor for analysis i and radionuclide j to convert a source into an intruder dose
$D_{i,j}$	=	the resultant dose to the intruder for analysis i from radionuclide j

The analysis index is either the Part 61 waste classification analysis (61) or the new analysis for incidental waste (N). Equation (B–1) can be written for both the old Part 61 waste classification analysis and the new analysis for incidental waste, and the two equations can be combined as Eq. (B–2)

$$\frac{C_{N,j}}{C_{61,j}} * \frac{V_N}{V_{61}} * \frac{X_{N,j}}{X_{61,j}} = \frac{D_{N,j}}{D_{61,j}} \qquad \text{(B–2)}$$

The conversion factor $X_{i,j}$ essentially represents the dose analysis to convert a quantity and distribution of radionuclide j into dose for analysis i. Many variables go into the conversion factor $X_{i,j}$, including but not limited to dosimetry, parameters, uncertainty treatment, and assumptions. The ratio of $X_{N,j}$ to $X_{61,j}$ in equation B-2 would look something like

$$\frac{X_{N,j}}{X_{61,j}} = \left(\frac{Dosimetry_{N,j}}{Dosimetry_{61,j}}\right) * \left(\frac{Parameters_{N,j}}{Parameters_{61,j}}\right) * \left(\frac{Uncertainty_{N,j}}{Uncertainty_{61,j}}\right) * \left(\frac{Assumptions_{N,j}}{Assumptions_{61,j}}\right) \qquad \text{(B–3)}$$

Although the Part 61 analysis could be duplicated to define the variables influencing $X_{61,j}$, the effort required would be substantial. A simpler but valid approach was used to develop the example averaging expressions for staff to use as a review tool. If it can be assumed for the Part 61 analysis that waste at the Class C limit for each radionuclide j multiplied by a conversion factor X_j resulted in a dose of 5 mSv [500 mrem] for the excavation intruder, then all of the variables except a new constant $[1/V_{61} * X_{N,j}/X_{61,j}]$ can be fixed in Eq. (B–2) without knowing the exact contribution of each of the terms in (B–3). Equation (B–2) can be solved for the term $[1/V_{61} * X_{N,j}/X_{61,j}]$ to yield the constants used in the example averaging expressions provided in Section 3.5.1.1. To apply the analysis to a drilling scenario as opposed to an excavation scenario simply requires defining V_N with respect to the primary variables affecting drilled waste volume.

The steps to complete the calculations were as follows:

(1)   Assume the Class C concentration values provided in Tables 1 and 2 of 10 CFR 61.55 would result in a 5 mSv [500 mrem] dose from each radionuclide for a typical commercial low-level waste disposal facility and the excavation intruder scenario using ICRP–2 dosimetry (for internal exposures, external exposures were also considered), the assumptions, and the scenario parameters for the dose assessment of a commercial low-level waste facility.

(2)   Use unit concentrations of each radionuclide (concentrations are either volume based or mass based consistent with Tables 1 and 2 of 10 CFR 61.55) as input and calculate the mean dose impact using ICRP 26 and 30 dosimetry with probabilistic calculations for each of the four scenarios described in Section 3.5.1.1 (shallow waste with no intruder barrier, shallow waste with an intruder barrier, deep waste with no intruder barrier, deep waste with an intruder barrier). For the four scenarios, the conditions for the calculation are shown in Table 3-1.

(3)   Calculate a **Part 61 dose concentration ratio** defined as the inadvertent intruder dose limit {5 mSv [500 mrem]} divided by the Table 1 and 2 Class C concentration value. This ratio provides impact per unit concentration resulting from the parameters, assumed dilution, and dosimetry approach for the Part 61 analysis.

(4)   For each radionuclide for two receptor types (acute and chronic), two disruption processes (drilling and excavation), and two waste access times (100 years and 500 years), calculate a **new dose concentration ratio** defined as the mean dose result for each radionuclide from the probabilistic calculations mentioned in Step (2) divided by the unit concentrations used as input for those calculations. The mean dose result should be selected in the year of peak dose following the intrusion event. For most calculations, the peak occurs in the year of the intrusion event. This ratio provides the impact per unit concentration for the parameters, assumed dilution, dosimetry, and uncertainty in the new calculations. Choose the limiting value by receptor type (acute or chronic) for each of the four scenarios (drilling at 100 years, drilling at 500 years, excavation at 100 years, and excavation at 500 years).

(5)   The **new dose concentration ratio** and the **Part 61 concentration ratio** are input into Eq. B–2, and the equation is rearranged to solve for an appropriate constant (the **new dose concentration ratio** is $D_{N,j}/C_{N,j}$; the **Part 61 concentration ratio** is $D_{61,j}/C_{61,j}$; V_N is known for the calculation).

(6)   The scenario-specific **classification expression** is developed based on the constant defined in Step (5) that is based on the primary variables of interest for the scenario (e.g., waste volume or waste thickness and drilling depth). If drilling depth is fixed at a site, then waste thickness would be the primary variable for a deep waste scenario. The primary variables of interest were selected to be those variables that directly define the amount of mixing for a scenario. Other variables could be selected, if necessary.

   (a)   For probabilistic calculations, the **new dose concentration ratio** is a matrix of results by (radionuclide, scenario). Each element of the matrix is a distribution of results.

   (b)   The **classification expression** is developed by taking the mean result for the limiting long- and short-lived radionuclides (e.g., for the resident postdrilling scenario at 500 years, the limiting long-lived radionuclide is Np-237). The limiting long- and short-lived radionuclides were used to simplify the expressions (i.e., to avoid developing a vector of constants by radionuclide for the classification expressions).

   (c)   The **classification expressions** for deeper waste (drilling scenarios) were developed using Equation B-2 with V_N defined with appropriate drilling scenario parameters (e.g., waste thickness).

The classification expressions provide the same protection per unit of material exhumed as the concentration limits in Table 1 and 2 and the Part 61 analysis provides; that is, both are based on an intruder dose of 5 mSv [500 mrem].

The radiological assessment method to develop the example averaging expressions calculated total effective dose equivalent (TEDE) as the product of radionuclides concentrations in environmental media and pathway dose conversion factors (PDCFs). The PDCFs were derived similarly to the method of Kennedy and Strenge (1992) for converting residual contamination into TEDE. The concentrations in environmental media were estimated by first calculating mixing of a user-defined source for a given intruder scenario (e.g., intruder drilling chronic, intruder construction acute), then applying simple mass transfer models to estimate radionuclide concentrations in air, water, and soil to account for radionuclide decay including ingrowth of daughters, sorption, solubilities, hydrologic transport, and other mass transfer and partitioning processes. The pathways considered for the acute scenarios were inhalation, direct radiation, and inadvertent soil ingestion. In addition to the acute pathways, consumption of plants and animals were considered for the chronic pathways. Other pathways may need to be considered in the site-specific analysis (e.g., dependent on site conditions, the drinking water pathway may or may not be substantially delayed and the impacts smaller than the direct waste exposure pathways). To develop the benchmark expressions, the animal consumption pathway was not a dominant contributor for limiting short-lived (Cs-137) and long-lived (Np-237) radionuclides. In addition, the scenario was defined as a resident but not a resident-farmer, thereby making the animal pathway less credible. Therefore, the animal pathway was not implemented in the final calculations. For the excavation chronic scenario, the resident was assumed to grow food crops and was exposed through ingestion of the food crops. Plant/soil concentration ratios were used to estimate the amount of radioactivity taken into the plant from a unit activity in the soil. Four different plant parts were estimated separately: leafy, root, fruit, and grain. Parameter uncertainty was explicitly considered in many of the inputs. Data used in the analysis were consistent with previous assessments of Kennedy and Strenge (1992) and

B–3

regulatory products (e.g., RESRAD, D&D). The dynamic simulation software package GoldSim® was used to perform the analysis.

Probabilistic analysis was used in the development of the example averaging expressions to account for variability and uncertainty (e.g., uncertainty in soil-to-plant transfer factors). A probabilistic assessment is preferred, but it is not required to develop site-specific (Category 3) averaging expressions. Parameters that should be varied or evaluated with sensitivity analysis are determined on a case-by-case basis. In the analysis used to develop the example averaging expressions provided in Section 3.5.1.1, the new dose concentration ratio, regardless of radionuclide, was sensitive to drilling scenario related parameters such as drilling depth, waste thickness, garden area, and drill cuttings distribution area. The new dose concentration ratio for excavation scenarios, regardless of radionuclide, was sensitive to the excavation-related parameters such as waste volume in the excavation and excavation volume. Parameters that directly define the concentration of waste in the environment after disruption can confidently be assumed to be important. For this analysis, the short-lived radionuclides were limited by Cs-137, and the long-lived radionuclides were limited by Np-237. Direct radiation exposure and plant consumption were the primary pathways for Cs-137, with inhalation and inadvertent soil ingestion secondary pathways. The doses from these pathways are sensitive to exposure times, consumption rates, soil-to-plant transfer factors, dose conversion factors, and transmission factors. For Np-237, the plant pathway and external radiation exposure were the primary pathways, with inhalation and inadvertent soil ingestion providing less than 10 percent of the total dose. Sensitive parameters were similar to those for Cs-137.

References

ICRP. "Report of ICRP Committee II on Permissible Dose for Internal Radiation." International Commission on Radiological Protection. 1959.

U.S. Nuclear Regulatory Commission (NRC). "Residual Radioactive Contamination From Decommissioning." NUREG/CR–5512. October 1992.

APPENDIX C
PUBLIC COMMENTS, ADVISORY COMMITTEE ON NUCLEAR WASTE COMMENTS AND RESPONSES

NUREG–1854, "Standard Review Plan for Activities Related to U.S. Department of Energy Waste Determinations," was published for interim use and comment in May 2006. NRC received public comments on the document, and they were accepted until July 31, 2006. NRC received 12 comment letters during the public comment period, which are addressed in this appendix. Comments were received from four State agencies, two public interest groups, and five individuals. In addition to the public comments, the staff also received comments and recommendations from the NRC Advisory Committee on Nuclear Waste (ACNW). The ACNW comments and recommendations are also addressed in this appendix.

NRC staff reviewed each of the comments received. Some of the comments were very similar to other comments. Similar comments were grouped and were given a single response. Several comments were outside the scope of NUREG–1854.

<u>Public Comments</u>

COMMENT: One commenter suggested that the guidance document should address climate change induced by increased carbon emissions.

RESPONSE: The review guidance directs NRC staff to consider the impact of short-duration, large magnitude events. Emphasis has been added in Sections 4.3.1 and 7.1.1.2 for those cases where fluvial erosion, flooding, slope failure, or landsliding may be significant. For most deeply buried waste systems, the impact of human-induced climate change is mainly expected to affect the timing of the change in climatic conditions at the waste but not the magnitude of the conditions, due to the damping function of the overlying geologic system. Temperatures and precipitation could increase or decrease depending on the particular site (e.g., local climates can be affected opposite to the global climatic trend). For wetter conditions, increasing infiltration rates using higher moisture contents and providing shallower water tables (at times earlier than would be estimated by consideration of the natural cycling of climate) would be an appropriate mechanism to test the sensitivity of the assessment potential doses or site stability to human-induced climate change. As the commenter noted, the main exception may be for a waste disposal system expected to be affected by surface geologic processes such as fluvial erosion, slope failure, or landsliding. Assessment of long-term surface geologic processes has many uncertainties, only one of which is the effect of climate change. If the assessment identifies a large sensitivity of system performance to uncertainty in the effects of climate change, bounding assumptions or the use of designs that mitigate the impacts should be evaluated. Guidance Sections 1.1.3.3, 4.3.1, 7.1, and 7.2 have been revised to address uncertainties in climate projections.

Predictive abilities for future climates continue to evolve. NRC staff involved in waste determination reviews will continue to remain abreast of developments regarding climate change, incorporating new information as appropriate.

COMMENT: One commenter suggested that NRC should reference the 2000 waste determination review for the Savannah River Site in Section 1.3 of the guidance document.

RESPONSE: The two prior waste determination reviews listed in Section 1.3 of the guidance document were included only as examples. There was no intent to provide an exhaustive list of prior waste determination reviews.

COMMENT: One commenter suggested that DOE's draft waste determination, its supporting technical basis documents, and any relevant NRC-generated documents should be made publicly available on NRC's website at the earliest possible time. The commenter also recommended that relevant documents should be made available prior to a draft comment period for any Technical Evaluation Report (TER) NRC staff generates to document its review of a waste determination.

RESPONSE: As described in Section 9.4 of the draft guidance, NRC staff will make relevant documents generated by NRC or submitted by DOE available on its website, with the possible exception of documents withheld because of security concerns. In practice, NRC staff publishes externally generated public documents to its Agencywide Documents Access and Management System (ADAMS) as soon as practical after it receives them. NRC staff typically publishes NRC-generated public documents to ADAMS within five days. NRC does not intend to publish draft versions of its TERs for public comment. Although the TER is not published for public comment, NRC does make its request for additional information public and conducts public meetings on specific waste determinations to assure the staff's review process is as transparent as possible. No changes have been made to the guidance in response to this comment.

COMMENT: One commenter agreed with the NRC that prior waste determinations should be reviewed and appropriately considered when reviewing DOE waste determinations at hand. The commenter also suggested that NRC analyze DOE waste determinations against National Academy of Sciences (NAS) recommendations. Furthermore, the commenter suggested that if a DOE waste determination was found not to be in full compliance with NAS recommendations, then the NRC "should not recommend going forward with the waste determination."

RESPONSE: The NRC Staff disagrees with the commenter's suggestion that the NRC should formally evaluate DOE waste determinations against the NAS recommendations. The staff believes that the NAS study (National Academy of Sciences, 2006) provides valuable recommendations, perspectives, and insights, but that it does not establish a set of criteria that must be met. NRC staff will continue to remain abreast of relevant NAS studies and the associated valuable insights.

COMMENT: Two commenters made suggestions on a number of aspects related to the dosimetry used in the performance assessment.

RESPONSE: The guidance document advocates the use of dosimetry consistent with 10 CFR Part 20. This ensures that compliance calculations for 10 CFR 61.41 and 10 CFR 61.43 remain consistent, as is discussed in NUREG–1573, "A Performance Assessment Methodology for Low-Level Radioactive Waste Disposal Facilities."

The proposed rule for 10 CFR Part 63, "Disposal of High-Level Radioactive Wastes in a Proposed Geological Repository at Yucca Mountain, Nevada; Proposed Rule," states

> "As a matter of policy, NRC considers 0.25 mSv [25 mrem] Total Effective Dose Equivalent (TEDE) as the appropriate dose limit within the range of potential

doses represented by the current 10 CFR 72.104 limit of 0.25 mSv [25 mrem] (whole body), 0.75 mSv [75 mrem] (thyroid dose), and 0.25 mSv [25 mrem] (to any other critical organ)." (64 FR 8644)

As 10 CFR 61.41 has the same standard as 10 CFR 72.104, this policy is applicable, and therefore, waste determinations will use TEDE without specifically considering individual organ doses. In addition, the guidance specifies that intruder calculations should also be based on 5 mSv [500 mrem] TEDE calculations without specifically considering individual organ doses, to be consistent with the dosimetry used for 10 CFR 61.41 and 10 CFR 61.43. Because of the tissue weighting factors and the magnitude of the TEDE limit, specific organ dose limits are not necessary to protect from deterministic effects.

Currently, NRC has a policy of allowing licensees to use these systems on a case-by-case basis if they request. While not identical, NRC would equate 0.25 mSv [25 mrem] effective dose the same as 0.25 mSv [25 mrem] total effective dose equivalent. In NUREG–1757, Volume 2, Appendix I.5.3.2, NRC addresses the issue for long-term performance assessment within the scope of decommissioning. The guidance states

"Licensees may request an exemption from Part 20 to use the latest dose conversion factors (e.g., International Commission on Radiation Protection [ICRP] Report 72). Scenarios and critical group assumptions should be revisited to look at age-based considerations. Licensees may not 'pick and choose' dosimetry methods for radionuclides (e.g., Federal Guidance Report No. 11 for six radionuclides and current International dose conversion factors for three radionuclides)."

DOE is not a licensee and would not require an exemption to use the alternate dosimetry. However, the guidance for WIR would be consistent with the guidance for decommissioning in that DOE should use the alternate dosimetry for all parts of the analysis (10 CFR 61.41, 61.43, and intruder analyses) and for all radionuclides. An exception may be for direct exposures, which were not updated for ICRP-72 and therefore would need to use the FGR-13 values. Additionally, staff would consider whether age-specific scenarios or exposure groups could result in higher doses.

COMMENT: One commenter expressed concern with the term "as DOE may authorize" in Section 2.4.4 as it related to waste that exceeds Class C concentration limits. The commenter suggested that this phrase might be construed to mean that DOE may call anything incidental waste and handle it any way that it "deems proper."

RESPONSE: The staff disagrees with this assertion and points out that in the same section of the guidance, the reviewer is directed to evaluate whether there is reasonable assurance that DOE alternate criteria for waste that exceeds Class C concentration limits are protective of public health and safety.

COMMENT: One commenter asked whether the term "comparable" meant that DOE could exceed the 10 CFR 61 limits by any amount DOE wished.

RESPONSE: Section 2.4.5 of the guidance explains that DOE could propose to dispose of waste that is not covered by the Ronald W. Reagan National Defense Authorization Act for Fiscal Year 2005 (NDAA) in accordance with alternate safety requirements. However, DOE is expected to provide an adequate technical basis that demonstrates that the proposed alternate safety requirements are "comparable" with the 10 CFR Part 61, Subpart C, performance objectives. To date, DOE has not proposed any criteria other than 10 CFR Part 61, Subpart C.

COMMENT: One commenter noted that an interim report about tank wastes at Savannah River Site (SRS) produced by the National Academy of Sciences (National Academy of Sciences, 2005) indicated that the authors had not seen analyses supporting DOE's claim that it would be impractical to dismantle and remove underground tanks used to store reprocessing waste.

RESPONSE: DOE's environmental impact statement for tank closure at SRS includes estimates of worker dose and financial costs that would be likely to result from exhumation of tanks at SRS, but does not include detailed supporting analyses. NRC staff recognizes that the estimated risks and costs associated with exhuming the tanks are uncertain because DOE has not previously performed similar activities.

Section 3.3 of the draft guidance directs reviewers to evaluate whether DOE considered technologies that minimize the volume of residual waste as well as technologies that selectively remove one or more highly radioactive radionuclides from the waste. In addition to these alternatives, it also is appropriate to consider the practicality of removing contaminated equipment for disposal instead of stabilizing it in place. Thus, it appears to be appropriate to consider the practicality of exhumation and disposal of waste tanks. NRC staff expects that the level of detail presented in any such analysis would be based on the level of detail necessary to distinguish exhumation from other options and may be limited by the uncertainty in available information. Text has been added to Sections 2.4.3 and 3.3 of the guidance to clarify this point.

COMMENT: A member of the public expressed the opinion that the $2,000 per person-rem conversion factor discussed in Section 3.4 of the draft guidance does not place an appropriate value on human life. The commenter expressed concern that the conversion factor will be devalued by inflation and suggested that an appropriate cost-benefit metric should be related to mortality data and appropriate discounting rates.

RESPONSE: NRC uses a $2,000 per person-rem conversion factor In regulatory analyses and some types of ALARA analyses. As described in NUREG–1530 (NRC, 1995), the factor is based on a representative value of a statistical life of $3 million multiplied by a risk coefficient of $7 \cdot 10^{-4}$ per rem. This risk coefficient accounts for the chance an exposed individual will experience nonfatal cancer, fatal cancer, or serious genetic effects, and a judgment of the severity of the nonlethal effects. The factor applies only to individuals receiving low doses of radiation that are below NRC's dose limits. It does not apply to deterministic health effects that could result from high doses to particular individuals, including early fatalities, because no death will ever be "acceptable" to NRC in the sense that NRC would regard it as a routine or permissible event. As discussed in NUREG–1530, NRC's dollar per person-rem conversion factor is revised periodically to account for changes in the value of the dollar (i.e., the effects of inflation), new research addressing the appropriate value of a statistical life, or changes in recommended radiation risk coefficients. In applying the conversion factor to future averted doses, NRC staff use economic discounting to adjust monetary values to present-day values.

Section 3.4.1 of the draft guidance describes reasons that NRC's $2,000 per person-rem conversion factor is not expected to be applicable to DOE waste determinations. In response to comments received about the applicability of collective dose metrics, this discussion has been expanded in the revised guidance.

COMMENT: Two commenters expressed opinions about the identification of highly radioactive radionuclides. One commenter suggested that NRC should ensure that long-lived mobile isotopes, such as Tc-99 and I-129, be designated as highly radioactive radionuclides. Another commenter observed that all radionuclides that contribute significantly to the dose to a potential receptor at any time during the period of assessment should be identified as highly radioactive radionuclides.

RESPONSE: As described in Section 3 of the draft guidance, NRC staff believes that highly radioactive radionuclides are those radionuclides that contribute most significantly to risk to the public, workers, and the environment. Because highly radioactive radionuclides are defined in terms of the risk they pose to various receptors, there is no single list of highly radioactive radionuclides that is used in all waste determination reviews. Rather, the choice of highly radioactive radionuclides is expected to be specific to each particular waste determination. NRC staff agrees with the commenter's observation that all radionuclides that contribute significantly to the dose to a potential receptor at any time during the period of assessment should be identified as highly radioactive radionuclides for the relevant waste determination. To clarify these points, an example has been added to Section 3.2 illustrating the selection of highly radioactive radionuclides when doses occur through different pathways and at different times.

The other commenter stated that NRC must ensure that adequate protection from long-lived mobile isotopes is achieved. NRC staff understands that the potential risks from long-lived, mobile radionuclides must be assessed carefully and believes that the review guidance provided in Sections 4 and 5 of the draft guidance will facilitate assessment of these risks. However, NRC staff notes that, as described in Section 10 of the guidance document, NRC's role is limited to assessing whether appropriate performance objectives will be met, rather than ensuring that appropriate performance objectives will be met, because NRC does not have regulatory authority over DOE's disposal actions.

COMMENT: Two commenters expressed opinions about the types of costs and benefits that should be included in the evaluation of radionuclide removal. One commenter expressed concern that the draft guidance does not direct reviewers to consider all relevant requirements or the full range of costs associated with the proposed disposal actions. As examples of additional requirements that should be considered, the commenter cited State laws, Federal laws such as the Comprehensive Environmental Response, Compensation, and Liability Act (CERCLA), Tribal treaty rights, requirements for the prevention of the spread of contamination, and requirements for the protection of groundwater for its best beneficial use. As an example of an additional cost that should be considered, the commenter cited costs associated with damage to natural resources and expressed the specific concern that the impact of uranium and technetium on groundwater at the Hanford site could be "immense." Another commenter expressed the view that scheduling and programmatic considerations should not be included in the assessment of whether highly radioactive radionuclides have been removed to the maximum extent practical. Specifically, the commenter expressed the opinion that such considerations are "beyond the bounds" of the fundamental principles of health physics, including the principle that doses should be maintained as low as is reasonably achievable (ALARA).

RESPONSE: NRC staff recognizes that additional State, Federal, and Tribal requirements may apply to the DOE at sites discussed in the draft guidance. The draft guidance does not direct reviewers to assess compliance with these additional requirements, because they are beyond the scope of NRC's authority. NRC staff understands that compliance with these regulations and obligations is likely to impose costs on DOE. However, because DOE must comply with applicable regulations and treaty obligations no matter which radionuclide removal activities are performed, the cost of complying with these obligations would not affect the analysis of the practicality of performing additional radionuclide removal. Therefore, the guidance has not been changed to direct reviewers to consider the costs of complying with these additional requirements.

The commenter's suggestion that the guidance should direct reviewers to consider costs associated with damage to natural resources is consistent with the guidance document review procedure that directs reviewers to consider reduction in risk or dose to the public, workers, and the environment as benefits in any cost-benefit analysis that supports a decision about the practicality of additional radionuclide removal. To clarify this point, a table listing reduction in impacts on natural resources as a potential benefit of additional radionuclide removal has been added to Section 3.4 of the guidance document. Similarly, the potential environmental impact of additional radionuclide removal activities (e.g., disruption due to exhumation of buried piping) is included as a potential cost of additional radionuclide removal.

The dose standards set in 10 CFR Part 61, Subpart C are protective of public health and safety, therefore, NRC staff does not agree that waste that meets the performance objectives would result in "immense" costs due to groundwater contamination. However, it is possible that releases of waste that meet the performance objectives of 10 CFR Part 61, Subpart C, could cause wellhead concentrations of some radionuclides to exceed U.S. Environmental Protection Agency's (EPA's) Maximum Contaminant Levels. As discussed in other contexts [e.g., NRC's guidance regarding ALARA analyses for license termination (NUREG–1757, Appendix N)], it is appropriate to include the potential cost reductions that other entities, such as public water supply utilities, may incur to meet the requirements of the Safe Drinking Water Act as benefits of additional radionuclide removal in cost-benefit analyses. Therefore, text has been added to Section 3.4 to explain this point, and potential cost reductions of other entities have been included in the new table of potential costs and benefits added to Section 3.4.

Although it can be appropriate to consider potential environmental impacts, NRC staff has not changed its guidance that the primary costs and benefits of additional radionuclide removal are expected to be financial costs and changes in the potential doses to workers and members of the public. Doses to workers and members of the public, as well as financial costs, are expected to be the primary considerations because waste that meets the performance objectives is expected to be protective of public health and safety. NRC staff recognizes that the relevant sites may have significant existing environmental contamination due to historical practices or accidents. However, remediation of this contamination is beyond the scope of the guidance, which pertains only to waste determined to be non-high level waste (HLW) pursuant to the NDAA or other applicable criteria, as discussed in Section 2 of the guidance document.

As described in Section 2.4 of the guidance document, NRC staff has concluded that the unqualified use of the word "practical" in the NDAA Criterion 2 allows for a relatively broad range of topics to be considered. NRC staff disagrees with the opinion that it is inappropriate to consider impacts on DOE's mission or schedule in assessing compliance with NDAA Criterion 2, because such impacts affect the practicality of radionuclide removal activities.

Furthermore, Section 3.4 directs reviewers to consider scheduling or programmatic considerations in terms of potential dose to workers or members of the public and financial costs. Changes in potential dose and financial costs attributable to schedule or mission impacts appear to be no less applicable to the assessment of compliance with NDAA Criterion 2 than any other financial costs or effects on dose associated with radionuclide removal. NRC staff also disagrees with the commenter's statement that such considerations are beyond the bounds of the ALARA principle and note NRC's definition of the ALARA principle explicitly includes "societal and socioeconomic considerations" in addition to technical and economic considerations (see 10 CFR 20.1003). For these reasons, no changes have been made to the guidance document in response to this comment.

COMMENT: One commenter noted that differences between the commercial disposal of low-level waste (LLW) and incidental waste make it inappropriate to apply the Class C concentration limits developed for commercial LLW to incidental waste.

RESPONSE: In response to this comment and other related comments, the guidance document sections related to concentration averaging have been revised. Details of the changes to the guidance are discussed in more detail later in this appendix.

COMMENT: One commenter expressed disagreement "with the NRC's characterization that DOE may implement Order 435.1's waste incidental to reprocessing (WIR) evaluation criteria to reclassify HLW in States not covered by the NDAA."

RESPONSE: The NDAA only applies to the States of South Carolina and Idaho. For waste determinations that DOE may make that are not covered under the requirements of the NDAA, the NRC has no role in a DOE decision to use DOE Order 435.1. The guidance document describes a systematic review process NRC staff will use to ensure consistency in its reviews of waste determinations, whether they are performed under the criteria specified in the NDAA or another set of criteria. Page xix of the guidance document has been clarified.

COMMENT: One commenter expressed concern that NRC may decide not to apply the same criteria to DOE's waste determination activities as to the New York State Energy Research and Development Authority's (NYSERDA's) license termination activities.

RESPONSE: The NRC's West Valley Policy Statement sets forth the criteria NRC staff will use to evaluate DOE's incidental waste determinations at the West Valley site. The Commission noted that it included WIR criteria in the Policy Statement "in light of the fact that the site will ultimately revert to control by NYSERDA under an NRC license [and so] both NYSERDA and NRC have an interest in ensuring that the incidental waste determination [made by DOE] need not be revisited." The Policy Statement further notes that "[c]onsistent with the overall approach in applying the [License Termination Rule] to the WVDP and to the entire NRC-licensed site following conclusion of the WVDP, the resulting calculated dose from the incidental waste is to be integrated with all the other calculated doses from material remaining ... at the entire NRC-licensed site."

COMMENT: A number of commenters expressed opinions on the ratio specified in the Category 2 approach for concentration averaging (factor of 10 for the ratio of unstabilized to stabilized radionuclide concentrations for waste classification purposes); some stated that it was too large and others that it was too small.

RESPONSE: The Category 2 approach to averaging is intended to be a risk-informed approach balancing the desire to consider aspects specific to waste classification for incidental waste with maintaining consistency with past principles while ensuring protection of public health and safety. 10 CFR Part 61 addresses stabilization of waste for disposal, but it is for the scenario of waste that can be well mixed with a stabilizing agent and placed in a container for shipment and disposal. The concentration averaging guidance for commercial low-level waste disposal addresses many common situations and different waste types.

The Category 2 approach is consistent with the principle expressed in 10 CFR Part 61 that averaging over the waste is permitted. Waste in incidental waste determinations may be bulk waste that can be mixed with a stabilizing agent, but it may also be thin layers of material that adhere to surfaces within large, engineered storage systems. The Category 2 approach is a simple and straightforward method of estimating concentrations for waste classification purposes without explicitly developing all of the constraining conditions and factors for every imaginable problem.

The stabilizing agent most commonly considered (to date) for incidental waste is cement or grout. A cementitious wasteform should readily be able to achieve a waste loading of 10 mass percent and probably higher. Therefore, the factor of 10 derived from this simple approach provides a reasonable boundary for determining when a straightforward waste classification argument can be made (based on the need to stabilize waste) and a more risk-based approach may be necessary (based on the need to stabilize waste in a particular disposal system). The factor of 10 is not arbitrary; rather, it is based on waste stabilization experience with cementitious materials.

A Category 3 approach has been developed to allow greater flexibility to consider the unique aspects of incidental waste classification. It is not an arbitrary mechanism to reclassify waste; rather, it is a risk-informed approach to recognizing the differences between incidental waste disposal scenarios and commercial low-level waste disposal scenarios. The greater flexibility comes with the expense of the increased burden of more detailed calculations and more review effort by NRC staff. The concentration averaging guidance has been revised in Section 3.5.1.1, and a new appendix (Appendix B) has been added describing the approach.

COMMENT: One commenter expressed the opinion that the types of concentration averaging approaches that are acceptable would not be protective.

RESPONSE: Concentration averaging that may be performed for waste classification purposes may be different from the approaches used to represent waste in a performance assessment. The guidance indicates that either the actual distribution of waste or a distribution that is conservative with respect to the estimated risk calculations needs to be considered in the performance assessment.

COMMENT: One commenter expressed concern that NRC may allow the disposal of wastes that do not meet the Class C concentration requirements.

RESPONSE: The NDAA allows DOE, in consultation with NRC, to dispose of waste that does not meet Class C concentration limits in South Carolina and Idaho, provided certain requirements are met. One of these requirements is that the performance objectives of 10 CFR Part 61, Subpart C, be met such that public health and safety is protected.

If NRC determines that a proposed approach would not likely be protective of public health and safety, regardless of the class of the waste, that conclusion would be made and documented in the relevant TER. Note that many additional factors (other than concentration) contribute directly to the acceptability of waste disposal, including depth of the waste, physical and chemical form of the waste, engineered barriers, natural barriers, and waste management practices.

COMMENT: One commenter suggested that the guidance sections related to radionuclide inventory, identification of highly radioactive radionuclides, and removal of highly radioactive radionuclides should include review procedures that direct reviewers to prepare concise statements documenting the uncertainties in DOE's analyses and NRC's conclusions about the adequacy of DOE's technical bases for its conclusions.

RESPONSE: NRC staff agrees that it is essential to document the main sources of uncertainty in analyses supporting waste determinations. Each set of review procedures the commenter highlighted includes procedures that direct reviewers to evaluate the uncertainty in DOE's analyses. The review procedures also provide guidance about the likely sources of uncertainty in each technical area.

NRC staff disagrees with the commenter's suggestion that the sets of review procedures are the appropriate places to provide guidance about review documentation. Instead, text has been added to Section 9 of the guidance, which provides guidance about review documentation to remind reviewers that a TER should include a description of the major sources of uncertainty in DOE's analyses. No change was made to the guidance document to indicate that reviewers should document their conclusions about the adequacy of DOE's technical bases, because Section 9 of the draft guidance indicates that a TER should document the staff's conclusions as to whether there is reasonable assurance that the applicable criteria can be met.

COMMENT: Two commenters expressed concerns about review guidance related to radionuclide removal activities that DOE plans to begin several years after a waste determination is submitted. One commenter suggested NRC should remove guidance that directs reviewers to evaluate DOE's plans to consider technological improvements that occur after submission of a waste determination if radionuclide removal activities are expected to begin 5 or more years later. The commenter expressed the opinion that it is unrealistic to expect that a new technology or method that could offer a measurable improvement in radionuclide removal without significantly increasing removal uncertainties could be introduced only 5 years after DOE performed its initial technology selection. The commenter also expressed the concern that continued consideration of technological improvements after submission of a waste determination could delay radionuclide removal and increase removal costs. Another commenter expressed the opposing opinion that NRC staff should refuse to review any waste determination that pertains to removal actions that would not occur for several years and recommended DOE should not submit a waste determination until it can "demonstrably prove" that its actions will comply with the NDAA provisions.

RESPONSE: NRC staff understands the concern that evaluating radionuclide removal technologies after submitting a waste determination could delay radionuclide removal or stabilization. However, NRC staff cannot conclude that highly radioactive radionuclides will be removed to the maximum extent practical if radionuclide removal technologies are selected significantly in advance of removal activities and no effort is made to consider whether application of improved technologies or methods could substantially improve radionuclide removal. NRC staff recognizes that the complete period of development of many radionuclide removal technologies may be longer than 5 years. However, NRC believes that 5 years is a reasonable amount of time for an existing technology that is considered to be technologically "immature" at the time of the waste determination to be further developed and tested so that its implementation would not significantly increase the uncertainty in predicted radionuclide removal.

NRC staff also understands the concern that the conclusions of a review performed prior to radionuclide removal are likely to be more uncertain than the conclusions of a review performed after radionuclide removal is complete. However, NRC staff disagrees with the view that NRC should refuse to review waste determinations submitted several years before radionuclide removal actions are implemented. NRC staff expects that, in most cases, waste determinations will be submitted less than 5 years before waste removal activities begin and that the guidance about technological developments that occur after submittal of a waste determination will apply to a small number of cases. However, in some cases, radionuclide removal plans may be sufficiently complex that some of the proposed removal actions must be planned several years in advance, and the removal actions may not be implemented for several years after the draft waste determination is submitted to the NRC for review (e.g., plans for salt waste disposal at SRS). In the TER documenting the review of a waste determination, NRC staff would discuss the major sources of uncertainty in DOE's conclusions, describe the key assumptions in DOE's analyses, and indicate whether there is reasonable assurance that DOE's assumptions are valid.

The commenter's specific recommendation that DOE should submit waste determinations only when it can prove that its actions will comply with the provisions of the NDAA does not appear to be applicable to determinations relevant to Hanford and West Valley, because the NDAA does not apply to those sites. NRC staff reviews waste determinations related to Hanford and West Valley only at DOE's request and believes it is appropriate for DOE to request a review whenever the review would be most useful to DOE. Although the conclusions of reviews performed at an early stage of planning are likely to be more uncertain than the conclusions of a review performed later, early interaction with DOE provides NRC staff an opportunity to assess DOE's disposal plans and, in its advisory role, recommend ways in which DOE's disposal actions or analyses may be made more robust.

NRC staff expects DOE to submit waste determinations relevant to SRS and Idaho National Laboratory (INL) when DOE concludes it has demonstrated the provisions of the NDAA will be met. NRC staff will then draw its own conclusions based on its review. If radionuclide removal activities have not been completed at the time of the review, NRC staff would identify the final radionuclide inventory as a key assumption to be verified during monitoring. In addition, NRC staff would evaluate the technical bases for the conclusion that the projected degree of radionuclide removal would be achieved. Although NRC staff will assess the validity of DOE's assumptions as part of its monitoring activities at SRS and INL, NRC staff bases its review conclusions on the data and analyses available at the time of its review. Therefore, significant uncertainty in the final inventory could prevent NRC staff from concluding that there is reasonable assurance that the performance objectives of 10 CFR Part 61, Subpart C, will be met.

Because NRC staff have concluded (1) it can be appropriate to review waste determinations that pertain to activities that will not be implemented for several years after the waste determination is submitted and (2) in those cases, it is appropriate for NRC staff to evaluate

DOE's program for monitoring developments in radionuclide removal technologies, no changes were made to the guidance in response to these comments.

COMMENT: A commenter identified a need for improved clarity in Example 2-2 of the concentration averaging section.

RESPONSE: The staff agreed with the comment, and the example in Section 3.5.1.1 has been clarified.

COMMENT: One commenter noted that, if radionuclide removal is not complete when a waste determination is reviewed, NRC staff should verify the final extent of radionuclide removal achieved. The commenter suggested that the need for verification should be stated explicitly in the sections of the guidance related to radionuclide removal (Section 3.3) and monitoring (Section 10).

RESPONSE: Section 3.3 of the draft guidance indicates that, if DOE submits a waste determination prior to ending removal actions, NRC staff will expect to monitor the extent of radionuclide removal achieved and assess any impacts of changes to the extent of radionuclide removal on meeting the performance objectives of 10 CFR Part 61, Subpart C. Section 3.3 has been clarified to indicate that NRC staff would expect to monitor the final radionuclide inventory only at SRS and INL. NRC staff would assess the impacts of any differences between the expected and actual extent of radionuclide removal achieved at Hanford only at DOE's request. As described in the guidance introduction, NRC's role with respect to license amendment or termination at West Valley is distinct from any consultation DOE may request regarding waste determinations relevant to the site. NRC staff expects to assess the final radionuclide inventory at West Valley at the time of license termination.

Section 3.3 of the draft guidance also indicates that, if removal activities are not completed as described in DOE's waste determination, NRC staff's conclusions about radionuclide removal to the maximum extent practical (NDAA Criterion 2) may no longer be applicable. The guidance does not indicate that NRC staff would reassess compliance with NDAA Criterion 2 based on the actual extent of radionuclide removal achieved, because the NDAA limits NRC's monitoring role to assessing compliance with the performance objectives of 10 CFR Part 61, Subpart C. However, as described in Section 4.7 of the guidance, compliance with the ALARA requirement of the performance objective for the protection of the general public from releases of radioactivity (10 CFR 61.41) is based in part on removal of highly radioactive radionuclides to the maximum extent practical. Thus, verification of the extent of radionuclide removal achieved would be included in NRC's monitoring role for SRS and INL in the context of the ALARA requirement of 10 CR 61.41.

Section 10 of the guidance describes the general framework of NRC monitoring pursuant to the NDAA. As indicated in Section 10, detailed instructions to NRC staff are provided in site-specific monitoring plans rather than in the guidance itself. However, Section 10 does indicate that NRC staff performing monitoring activities should verify DOE's key assumptions and lists the final inventory as an example of an assumption that is likely to be important to assessing compliance with the performance objectives of 10 CFR Part 61, Subpart C.

COMMENT: One commenter expressed an opinion on the inappropriateness of using deterministic analysis for incidental waste performance assessments and the difficulty in selecting conservative deterministic parameters.

RESPONSE: The staff agrees that certain types of problems are very challenging to evaluate with a deterministic performance assessment. As a generality, this would include complicated, highly uncertain problems that respond to independent variables in a highly nonlinear way. However, there may be cases where a deterministic analysis, with a comprehensive sensitivity analysis, can provide adequate basis to make a decision. In addition, there may be cases where a simple conservative calculation may be used to demonstrate compliance. The guidance document must apply to many different types of problems, and therefore the flexibility to use different approaches has been maintained. To ensure that an appropriate analysis approach has been used in the performance assessment, specific review procedures were provided in Section 4.4.2. These review procedures have been enhanced in the revised guidance. Note that when appropriate, NRC staff will perform independent probabilistic analysis to enhance the review of DOE analysis. Independent analysis helps the staff to better evaluate the DOE evaluation of uncertainty.

COMMENT: One commenter suggested the need to enhance the review procedures related to flow paths for transport to include manmade connections.

RESPONSE: NRC Staff agrees with this comment and has revised the review procedures in Section 4.3.4.2.

COMMENT: One commenter suggested that the review procedures related to the presence of fast pathways through aquitards needed to be enhanced.

RESPONSE: NRC Staff agrees with this comment and has enhanced the relevant review procedure in Section 4.3.4.

COMMENT: One commenter suggested that enhancing the review procedures in Section 4.3 was needed to add the consideration of past, present, or future plumes of organic solvents that may affect radionuclide transport.

RESPONSE: A review procedure addressing organic solvents has been added to Section 4.3.4. Adequate model support is needed for radionuclide transport projections. If model support is adequate, it should be unlikely that organic solvents may have modified the vadose zone or an aquifer such that transport times are greatly overestimated. Nonetheless, modification of the natural system from past site uses should be considered, particularly if radionuclide transport in natural materials significantly reduces risk.

COMMENT: Two commenters recommended changes to NRC's guidance that, in most cases, a buffer zone is expected to extend approximately 100 m [330 ft] from the disposal area. One commenter recommended that the guidance should indicate that a buffer zone should never extend more than 30 m [100 ft] from the waste and should explicitly state that a buffer zone should not allow for dilution to achieve compliance, including compliance with applicable drinking water standards. The commenter also expressed the view that institutional controls should not be assumed to last for more than 100 years. Another commenter expressed an opposing view and suggested that buffer zones for DOE non-HLW sites should typically be larger than buffer zones for commercial LLW sites to account for DOE facilities near reprocessing wastes (e.g., production reactors or chemical reprocessing facilities near tank farms). The commenter also noted that NRC had previously accepted DOE's use of a buffer zone larger than 100 m [330 ft] in a waste determination review performed at DOE's request (NRC, 2000).

RESPONSE: The 100 m [330 ft] buffer zone size recommended in the draft guidance is intended to be consistent with the purpose of a buffer zone as described in 10 CFR 61.7(a)(2), which is to provide a controlled area to establish monitoring locations and to take mitigative measures if needed. When drafting 10 CFR Part 61, NRC considered requiring that buffer zones extend a minimum of 30 m [100 ft] from the waste to allow sufficient space for monitoring and mitigative measures, if necessary. In comments on the draft rule, several individuals recommended that this minimum requirement be removed and that appropriate buffer zone sizes be evaluated on a case-by-case basis. In addition, the Army Corps of Engineers noted a 30 m [100 ft] buffer zone may not allow sufficient space for taking remedial action (NRC, 1982). Because NRC staff originally had intended to evaluate the appropriate size of a buffer zone on a site-specific basis, NRC decided not to include a specific size requirement for buffer zones in 10 CFR Part 61 (NRC, 1982). NRC staff has not changed its view that it is appropriate to consider the appropriate size of a buffer zone on a case-by-case basis and therefore did not alter the draft guidance to indicate buffer zones should never extend more than 30 m [100 ft] from the waste. Instead, text was added to Section 4.1.1.4 to remind reviewers of the purpose of a buffer zone.

NRC staff believes that, in most cases, a 100 m [330 ft] buffer zone will allow sufficient space for performing monitoring activities and mitigative measures (if necessary). Although buffer zones larger than 100 m [330 ft] may be acceptable, in general, buffer zones should not be enlarged solely to increase radionuclide dilution. NRC staff agrees with the commenters' suggestion that neighboring facilities may necessitate a buffer zone greater than 100 m [330 ft], but only if it is demonstrated that the neighboring facilities would prevent appropriate monitoring programs or any necessary mitigative actions from being implemented in a 100 m [330 ft] buffer zone. The commenters' suggestion that a buffer zone should be extended "based on source terms in the vicinity of the tanks such that the public should not have access" appears to imply that a buffer zone could cover any area that DOE intends to control to prevent public access. Thus the merit of the suggestion depends on what assumption NRC should make about the amount of time DOE should be required to maintain control of the land.

As described in the draft guidance, NRC staff assumes that DOE will actively control public access for 100 years after disposal site closure. NRC staff has previously considered whether the 100-year period of institutional controls specified in 10 CFR Part 61 should be extended (NRC, 1982). Although considered in the context of commercial LLW sites, the previous deliberations are relevant to waste determination discussions because 10 CFR Part 61 requires that, after closure, control of commercial LLW sites be transferred to the Federal or a State government [10 CFR 61.59(a) and 10 CFR 61.59(b)]. A number of individuals commenting on the draft version of 10 CFR Part 61 suggested that the period of institutional controls should be extended because the Federal or a State government could be expected to survive for much longer than 100 years. However, the 100-year limit is not based on how long the government is expected to survive, but on how long the government should be expected to provide active custodial care of the waste. Furthermore, intrusion into waste is not expected to result from the demise of the government, but rather from an error (NRC, 1982). In regional workshops and written public comments on a draft version of 10 CFR Part 61, a consensus was developed supporting the 100-year limit. Based on these and other considerations, NRC staff concluded that 100 years is an appropriate limit for reliance on institutional controls.

Because many radionuclides typical of reprocessing waste have half-lives of several thousand years or more, in some cases the peak dose to a potential receptor is predicted to occur several thousand years after site closure. Thus, any suggestion that the size of a buffer zone should be based on DOE's institutional control of the land would need to anticipate governmental control of the land for several thousand years and provide justification that the institutional controls would be durable throughout the period. Therefore, the guidance has not been changed to indicate that DOE non-HLW disposal sites generally should have larger buffer zones than commercial LLW disposal sites.

The commenter is correct in noting that, in a review of a waste determination performed in 2000, NRC staff agreed that it was reasonable for DOE to postulate a buffer zone larger than 100 m [330 ft]. That review was performed at DOE's request prior to promulgation of the NDAA. Because NRC was advising DOE at DOE's request, NRC performed an analysis based on the buffer zone size agreed to by DOE and the South Carolina Department of Health and Environmental Control, which had regulatory authority over aspects of DOE's tank closure plan. During the review, NRC staff requested that DOE evaluate predicted doses to a potential receptor closer to the tank farm, including 100 m [330 ft] from the tanks, to assess the sensitivity of dose to receptor location. Since that time, NRC performed an additional review of a waste determination for SRS (NRC, 2005) and three reviews of waste determinations for INL (NRC, 2002, 2003, 2006). In all of these determinations, DOE used a 100 m [330 ft] buffer zone, which NRC staff concluded was appropriate.

The commenter also correctly noted that, in some cases, the point of maximum exposure for the public may be more than 100 m [330 ft] from the disposal area. For example, a complex hydrogeologic system or overlap of plumes from multiple sources may cause peak concentrations of radionuclides in groundwater to occur more than 100 m [330 ft] from the disposal area. In a recent review of a waste determination for tank closure at INL (NRC, 2006), NRC staff agreed that it was appropriate to assume that a member of the public was located 600 m [200 ft] from the tank farm because that was the point DOE expected would yield the maximum greatest dose (for evaluation of 10 CFR 61.41). In this case, DOE assumed the buffer zone extended 100 m [330 ft] from the disposal area. To clarify this point, text has been added to Section 4.1.1.4 of the guidance to indicate that, to evaluate compliance with 10 CFR 61.41, a public receptor should be assumed to be located at the point of maximum exposure beyond the buffer zone, and the point of maximum exposure may not coincide with the boundary of the buffer zone.

NRC staff agrees that a buffer zone should not be enlarged solely to allow for radionuclide dilution and has amended Section 4.1.1.4 to make this point. However, NRC staff notes that it does not assess compliance with drinking water standards. As discussed in the guidance document introduction, NRC's role is limited to assessing compliance with the performance objectives of 10 CFR Part 61, or potentially, for West Valley and Hanford, performance objectives comparable to the performance objectives of 10 CFR Part 61. NRC staff recognizes that EPA or the affected States may implement regulatory programs for groundwater protection and expects that DOE and the affected parties will agree on any buffer zones that are relevant to the applicable regulations. These buffer zones may not coincide with buffer zones NRC and DOE used for waste determinations.

COMMENT: One commenter asked whether DOE's standards for the protection of public health and safety would be equivalent to NRC's standards.

RESPONSE: As described in Section 2 of the draft guidance, the waste determination criteria relevant to Hanford and West Valley are similar but not identical to the criteria that apply to SRS and INL. At SRS and INL, the applicable criteria (established in Section 3116 of the NDAA) require that waste meets NRC's performance objectives for the land disposal of radioactive waste (10 CFR Part 61, Subpart C). Thus, in waste determinations relevant to SRS and INL, there would be no difference between the DOE-applied standards and NRC's performance objectives.

DOE Order 435.1 and the West Valley Policy Statement indicate that waste may meet requirements that are "comparable to" NRC's performance objectives. Thus, in waste determinations relevant to Hanford and West Valley, there may be differences between the DOE-applied standards and NRC's performance objectives. If DOE proposes to apply requirements different from the performance objectives of 10 CFR Part 61, Subpart C, in a waste determination and if DOE consults with NRC about the determination, NRC staff would expect to discuss any differences between DOE's proposed requirements and NRC's performance objectives in a TER. As described in the guidance introduction, NRC's role with respect to waste determinations at West Valley is governed by the West Valley Policy Statement and NRC's role with respect to waste determinations at Hanford is at DOE's request. Because these points are explained in the Introduction and Section 2 of the draft guidance, no changes have been made to the guidance document in response to this comment. It should be noted that to date, DOE has not proposed any criteria other than that found in 10 CFR Part 61.

COMMENT: One commenter asked what is the likelihood that waste will be isolated from the biosphere for the time that it is required to be isolated.

RESPONSE: The probability that waste will remain isolated, or what fraction of the waste is released and when, is estimated in the performance assessment calculations. With a properly designed waste disposal system, most of the radioactivity should decay in place prior to eventual degradation of the facility.

COMMENT: One commenter suggested that NRC should provide the models and all model parameters and assumptions to interested parties upon public request.

RESPONSE: The NRC documents its review in requests for information and TERs, which are available to the public. To risk inform the review, NRC staff may perform independent technical analyses and develop computational models. The analyses and models are a review tool to inform the evaluation and are not the basis for conclusions. Conclusions of a waste determination review are based on DOE analyses and models and their associated results. NRC does not intend to make internal review tools publicly available. In the course of a review, NRC staff attempts to evaluate any information relevant to the technical review, including information from non-DOE sources. NRC staff willingly considers analysis performed by other stakeholders.

COMMENT: One commenter suggested that there is little evidence that institutional controls can be maintained for periods even approaching 100 years.

RESPONSE: Consistent with 10 CFR Part 61 for commercial LLW disposal, active institutional controls are assumed to be maintained for 100 years following closure of an incidental waste system. The only direct information to support the validity of the assumption is the Beatty, Nevada LLW facility that has been successfully closed, monitored, and maintained since 1992. Operating LLW facilities have not had inadvertent uses of their sites. DOE sites have maintained both active and passive institutional controls for 50 years or more. For a commercial LLW facility, there are requirements of the land owner or custodial agency with respect to institutional controls, which provide the primary basis for assuming a 100-year effectiveness period. DOE has requirements for control and maintenance of its sites.

COMMENT: One commenter pointed out that 10 CFR 61.41 specifies the dose limit to "any member of the public." The commenter suggested that the guidance document should provide additional guidance for either deterministic or probabilistic assessments as to the desired endpoint and confidence bounds for analysis.

RESPONSE: Section 4.3.5.1.2 discusses the average member of the critical group, which is the receptor definition NRC uses for dose assessments. A corresponding review procedure has been added to Section 4.3.5.2. Section 4.6.2 provides a review procedure to ensure that appropriate performance measures have been used in the performance assessment. For a deterministic analysis, it is peak dose. For a probabilistic analysis, it is peak mean dose. For most probabilistic assessments, the mean dose is a high percentile of the resulting dose distribution.

COMMENT: One commenter stated that NUREG–1757 was incorrectly referenced as requirements in Section 8, that the requirements of 10 CFR Part 50, Appendix B, do not apply, and that graded quality assurance should be followed.

RESPONSE: The incorrect reference in Section 8 has been changed. The reference to 10 CFR Part 50, Appendix B, was only provided as an example of an acceptable quality assurance program to ensure data quality, and therefore, the reference is appropriate and has been maintained. The staff agrees that graded quality assurance may be used, and additional text has been added to Section 8.

COMMENT: Two commenters stated that the content of the draft monitoring section of the guidance document went beyond what was intended under the NDAA. The commenter suggested that monitoring should focus on measurements that indicate overall performance of the facility, similar to monitoring performed at other contaminated sites where environmental monitoring is emphasized. Furthermore, the commenters suggested that monitoring to substantiate DOE's assumptions and analyses is impractical in an annual monitoring cycle, while reviewing site monitoring reports and other environmental reports is very appropriate and consistent with NRC practice.

RESPONSE: Section 10 (Monitoring) of the guidance document has been extensively revised and now discusses how NRC interprets its role in monitoring disposal actions under the NDAA and how this differentiates from environmental monitoring. The new section on overall approach and scope of NDAA compliance monitoring in the guidance document states that the NDAA indicates NRC must monitor DOE's disposal actions to assess whether the actions are in compliance with the performance objectives of 10 CFR Part 61, Subpart C. Because the performance objectives include a general requirement that there is reasonable assurance that exposures to humans will not exceed the limits established in 10 CFR 61.41 through 44, DOE's disposal actions would be out of compliance if NRC no longer had reasonable assurance that the requirements of 10 CFR 61.41 through 44 would be met. As described in Sections 4–7 of this guidance document, NRC staff typically derives assurance that the requirements of 10 CFR 61.41 through 44 will be met on predictions of long-term site performance. Therefore, monitoring to assess compliance with the performance objectives is expected to include activities necessary to maintain confidence in DOE's predictions of long-term site performance. Although environmental monitoring (i.e., monitoring the environment for changes and collecting environmental data) is expected to provide the most clear and straightforward way to assess the performance of a disposal facility, in practice there are limitations to solely relying on environmental data to evaluate the performance of the type of disposal facilities that are the subject of DOE's waste determinations. Because DOE typically relies on a number of engineered features (e.g., grout, concrete vaults, specialized covers) to close their facilities, it may be several decades or centuries before any radioactive materials are expected to be released from the disposal facility, and thereafter be detected. While observing DOE actions and reviewing environmental data are expected to be components of NRC's monitoring activities, building confidence in DOE's selection of parameters and models will be a critical monitoring activity. Accordingly, it is important for staff to use the performance assessment to identify those aspects of the disposal system that have the largest influence on performance so that these areas can be the focus of NRC's monitoring activities. In addition, it is important for staff to monitor those parts of the disposal system with here there is the greatest amount of uncertainty that NRC's performance objectives will be met.

COMMENT: Two commenters suggested that the West Valley site should be included in the discussion of monitoring in the guidance document. One of the commenters pointed out that the West Valley Demonstration Project Act calls for NRC to monitor DOE's activities at West Valley for the purpose of assuring public health and safety and believes that this is similar to NRC's role under the NDAA. The commenter believes that NRC's monitoring at West Valley is a good example of how monitoring can and should be performed and should provide a benchmark to measuring monitoring proposals under the NDAA.

RESPONSE: NRC staff believes that there are important differences between its monitoring role under the NDAA and at West Valley. NRC's monitoring role under the NDAA is narrowly focused on assessing DOE's compliance with the performance objectives set out in 10 CFR Part 61, Subpart C. As explained in the guidance, a large part of this responsibility is predictive in nature in that NRC must assess whether it continues to have reasonable assurance that the performance objectives will be met in the future. NRC's role at West Valley is broader in that NRC monitors all of DOE's activities under the demonstration project and the focus is more on ongoing activities such as those associated with the environmental monitoring program, disposal area maintenance, and review of reports and analyses. NRC examines the future status of the West Valley site by participating in the preparation of the environmental impact statement and by verifying DOE's compliance with NRC's decommissioning criteria. At this point, no WIR waste determination process has been started at West Valley. If appropriate, guidance on monitoring disposal actions pursuant to a waste determination at West Valley may be included in a revised version of this guidance document.

COMMENT: Two commenters questioned how NRC staff planned to monitor different aspects of disposal actions (e.g., how would NRC staff monitor infiltration rates and wasteform degradation?). In addition, one of the commenters suggested that certain monitoring activities are unrealistic or unnecessary. The examples provided were (1) monitoring intrusion performance would not be pertinent until well beyond the 100-year institutional control period (2) protection of individuals during operation and offsite releases are well covered by various DOE orders consistent with 10 CFR Part 20 and duplicate monitoring efforts are not needed; (3) monitoring each batch of waste and actual sampling of some or all the batches by NRC staff seems unreasonable and is not required; and (4) monitoring the performance assessment process and its updates is not normally part of the monitoring program.

RESPONSE: To confirm key aspects of or assumptions used in DOE's performance assessment to demonstrate compliance, NRC staff will carry out technical review and onsite observation activities. If infiltration rates and wasteform degradation were identified as being key aspects in NRC's technical evaluation, monitoring activities could include observing relevant experiments, reviewing the results of experiments, and reviewing analyses or expert elicitation regarding the long-term performance of the infiltration barrier or wasteform. Specific examples of technical information evaluated for monitoring of infiltration and wasteform performance could include reviewing moisture characteristics for the grout, curing time and techniques for the grout, and information on the dissolution of salts and low solubility matrix phases within the grout.

To assess the potential for noncompliance in the future for a particular performance objective (e.g., 10 CFR 61.42), key aspects of the performance demonstration should be monitored in the present. This could include monitoring activities that substantiate which erosion control designs adequately prevent inadvertent intrusion and onsite observations of the construction of an erosion barrier. Potential long-term compliance monitoring activities to substantiate key aspects of DOE intruder analyses may include assessing erosion cover maintenance including maintenance to control subsidence, biotic intrusion, and vegetative cover performance.

Monitoring activities to assess compliance with 10 CFR 61.43 are not expected to duplicate DOE efforts, but instead will help to understand and assess DOE's monitoring program. This will be done in a graded approach that will require fewer NRC monitoring activities as the monitoring period continues.

NRC staff plans to observe waste sampling activities, review waste sampling data, review DOE's waste sampling plans, and review quality assurance procedures for waste sampling to substantiate stated radioactive inventory. If a key aspect of DOE's performance demonstration is that the highly radioactive radionuclide concentrations in the waste would not be greater than the projected concentrations that were used to support DOE's waste determination, then NRC staff would request information about radionuclide concentrations. However, NRC staff does not plan to perform waste sampling, but instead is expected to observe and review DOE's waste sampling activities.

DOE's performance assessment is a critical element of its demonstration of compliance with the performance objectives for 10 CFR 61.41 and 42 because it forms the primary basis for concluding that the performance objectives will be met. Because the performance assessment is important to the performance demonstration, NRC staff must evaluate the adequacy of revisions and updates to it. The depth of the performance assessment review will depend on the type of changes to DOE's performance assessment and how these changes and the impacts of the changes are documented and referenced.

COMMENT: Three commenters correctly pointed out that the NDAA has no requirements for DOE to submit a monitoring plan to the NRC and that NRC staff should be cognizant that DOE will not develop a monitoring plan to solely meet the NDAA because it is an NRC responsibility.

RESPONSE: NRC is cognizant that it is NRC's responsibility to monitor DOE's disposal actions to assess compliance with the performance objectives set out in 10 CFR Part 61, Subpart C, and that, although DOE is required to develop a monitoring plan for DOE's internal use under DOE Order 435.1, it is NRC's responsibility to develop a monitoring plan to meet the requirement stipulated under NDAA, Section 3116, Paragraph (b)(1). Section 10 has been clarified on this point.

COMMENT: Two commenters pointed out that NRC had oversight for both the Hanford and Barnwell disposal sites until the 1990s. NRC inspectors and headquarters staff inspected both facilities each year and reviewed the annual monitoring data to assure that the site continued to operate in compliance with the 10 CFR Part 61, Subpart C, performance objectives. Further, the commenter suggested that the covered States and NRC could conduct similar environmental monitoring visits as often as necessary for NRC to fulfill its obligations under the NDAA.

RESPONSE: For most facilities, DOE currently relies on robust engineered barriers to isolate waste and contaminants, and it may be several decades or centuries before radioactive materials are expected to be released from the disposal facility. In comparison, the Hanford and Barnwell disposal sites rely on relatively simply designed facilities with fewer engineered barriers, unlike the potential non-high-level waste disposal sites covered under the NDAA. Although NRC relied more on environmental data and the activities of the State inspectors for monitoring information during this time period, NRC was engaged and active in long-term performance issues, as demonstrated by its examination and review of various issues by the Office of Research. For example, the Office of Research performed research on the use of engineered surface covers to enhance waste isolation over longer periods. Therefore, NRC intends to use both onsite observation as well as review of any DOE research and other activities related to long-term performance to ensure there will be reasonable assurance that the performance objectives will be met.

COMMENT: Two commenters suggested that there was a need for dialogue and possible resolution of compliance issues before noncompliance notification letters are sent. The commenters further stated that DOE would expect to resolve compliance issues before such notifications were needed. Other comments stated that draft noncompliance notification letters should be provided to DOE, and that comments from DOE and the covered State on the compliance issues be released together with the final noncompliance notification letter.

RESPONSE: Section 10 of the guidance document has been revised. A graded approach is described in which DOE is given the opportunity to understand NRC's concerns with respect to compliance issues and DOE is allowed time to identify an appropriate path forward and to respond. The guidance document section on noncompliance notification letters states that

> "Prior to sending out Type I–III letters, NRC will review its concerns in a letter (Type IV) to DOE and the State. This will give the State an opportunity to provide input and comment and provide DOE with an opportunity to provide information that demonstrates its disposal actions are in compliance with the performance objectives. Assuming that DOE provides information to support its performance demonstration, NRC will review this information and decide whether it is sufficient to conclude that there is reasonable assurance that the performance objectives will be met. If the staff determines that, based on the information DOE provided, there is sufficient basis to conclude that DOE is in compliance, NRC will send out a notification of resolution letter (Type V). Type IV and V letters will be made publicly available on NRC's website. These letters formally document resolution of the issue. If the staff determines that, based on the information DOE provides,

there is still a basis for concluding that DOE is noncompliant, NRC will transmit the noncompliance notification letter (i.e., Type I-III)."

In addition, the periodic compliance monitoring report should also serve as a vehicle for tracking the status of the monitoring activities in terms of whether the activity is considered closed, open, or open-noncompliant. Before the periodic compliance monitoring report would be published, it is envisioned that staff will regularly meet with DOE and the State to discuss the status of the monitoring program. This meeting could be especially useful in discussing activities that are still open, the actions that are needed to close activities, and potential areas of concern.

The NDAA does not require NRC and NRC does not intend to provide DOE with a draft periodic noncompliance notification letter, nor does NRC intend to simultaneously release potential DOE and State comments on the draft noncompliance notification letter with a final noncompliance notification letter. However, all types of notification letters (listed in Table 10-2 of the guidance document: noncompliance, concern, and resolution) and the periodic compliance monitoring report will be made publicly available and will be accessible on NRC's website.

COMMENT: Three commenters provided comments related to the covered State's role in NDAA compliance monitoring. One comment suggested that monitoring should not be modified to satisfy the requirements of the covered State's role unless NRC needs that information to carry out its responsibility. Another comment recommended that additional consultation should occur with the covered State to ensure that the essential components of monitoring under the NDAA are addressed in the monitoring chapter of the guidance document.

RESPONSE: NRC staff agrees that monitoring activities should not be modified to satisfy the requirements of the covered State's role unless NRC needs that information to carry out its responsibility under the NDAA. With regard to consultations with the covered state, the revised monitoring section of the guidance document states that

" ... it is recommended that discussions with the covered State concerning monitoring occur during the waste determination review. Discussions with the State will enable NRC staff to determine the type of activities the State will be undertaking in terms of monitoring compliance with its regulatory requirements, and enable the covered State to understand the type of activities that NRC will be undertaking. Insights gained through these discussions may help the two parties coordinate their activities in order for NRC to meet its NDAA monitoring responsibilities. In some cases, NRC may be able to rely upon State-obtained information."

COMMENT: Three commenters provided comments related to the covered State's role in determining what actions would be taken in response to a noncompliance notification letter. The commenters suggested that the State is not necessarily in a position to take action. One commenter suggested that it is possibly inappropriate to say that the covered State has specific regulatory authority at the DOE site and that it would be more appropriate to say that the covered State may have specific regulatory authority at the DOE site.

RESPONSE: Section 10 of the guidance document has been modified to clarify the covered State's role with regard to any potential regulatory authority.

COMMENT: One commenter suggested that the number of DOE site visits seemed inadequate and that local residents should be included in site visits.

RESPONSE: Section 10 of the guidance document has undergone extensive revision since the draft document was published. Section 10 now contains a complete discussion about the need for onsite observation visits and the resources needed when NRC staff goes to the site to monitor DOE waste disposal actions. The number of onsite visits needed will depend on the nature and number of monitoring areas and the status of DOE activities. As monitoring proceeds, new monitoring areas could be developed. However, it is expected that NRC's monitoring activities will decrease over time. DOE controls access to DOE sites, and NRC has a very limited role with regard to who should have access to the DOE site. Further, NRC will prepare reports after each site visit that documents the results of each visit. The reports will be publically available.

COMMENT: One commenter suggested that NRC include information in the guidance document regarding how DOE and NRC will comply with the National Environmental Policy Act (NEPA).

RESPONSE: Because NRC's role is limited to reviewing DOE actions, NRC is not required to prepare an Environmental Impact Statement under the NEPA requirements. DOE is responsible for fulfilling its own NEPA requirements, and the NRC has no role or authority over those activities. No changes were made to the guidance as a result of this comment.

COMMENT: One commenter suggested that the guidance document is based on the flawed assumptions that barriers will perform as designed for very long periods of time and that institutional controls will be maintained in perpetuity.

RESPONSE: The guidance document is not based on the assumption that institutional controls will be maintained in perpetuity; rather, it is assumed that institutional controls will be maintained for 100 years, consistent with the institutional control period for commercial LLW facilities. The document does provide review procedures to evaluate the performance of engineered and natural barriers. Reviewers are directed to consider relevant experience and uncertainties in the estimates of long-term barrier performance.

COMMENT: Two commenters suggested that complex geology can invalidate the assumption of plug flow in homogeneous materials.

RESPONSE: The review guidance does not presuppose that flow will be in homogenous, isotropic media. Rather, review procedures are provided to evaluate uncertainty in flow paths, particularly flow in preferential pathways.

COMMENT: Two commenters pointed out that isotopes of uranium, technetium, iodine, neptunium, americium, selenium, and cobalt form anions in soil and that the soils do not easily retain anions (with specific reference to the Hanford site).

RESPONSE: The guidance document provides review procedures for evaluating radionuclide sorption in a variety of materials, including natural materials, considering site-specific conditions. The staff agrees that the electrical charge on the chemical species can significantly affect transport phenomena.

COMMENT: One commenter pointed out the need to consider complexities when evaluating evapotranspiration (ET) barriers.

RESPONSE: The staff agrees that assessing ET covers must consider variation in processes at an appropriate temporal scale. New review procedures have been added in Section 4.3 to provide guidance specific to ET barriers.

COMMENT: Three commenters pointed out that the NDAA did not have any legal standing in the States of Washington and New York.

RESPONSE: On page xviii of the guidance document, it states that the NDAA is applicable only in the States of South Carolina and Idaho.

COMMENT: One commenter suggested that the guidance allows averaging of stabilizing materials and grout, while specifically directing that such averaging is not to be used in performance assessments.

RESPONSE: Averaging is allowed for waste classification because the exposure scenarios used to develop the classification limits would be expected to result in mixing of waste and nonwaste during waste exhumation. Averaging is also allowed in the performance assessment calculations if the assumption of waste distribution can be demonstrated to be conservative. If it cannot be demonstrated to be conservative, then the actual distribution of the waste in the system should be used to most accurately represent the release of waste into the environment. No changes to the guidance have been made to address this comment.

COMMENT: One commenter expressed concern that NRC's evaluation of compliance with the performance objectives of 10 CFR Part 61, Subpart C, will not provide adequate protection if the requirements of other parts of the regulation are not met.

RESPONSE: NRC staff understands the commenter's concern that 10 CFR Part 61 was intended to be applied in its entirety, but disagrees with the commenter's conclusion that NRC must evaluate compliance with all requirements of 10 CFR Part 61 to be sufficiently protective. NRC staff believes that compliance with the performance objectives of 10 CFR Part 61, Subpart C, will protect public health and safety. Furthermore, 10 CFR Part 61 was designed to be applied to commercial LLW disposal facilities. Consequently, many of the requirements of 10 CFR Part 61, such as Subpart E requirements for financial assurances, are not applicable to a Federal agency. With respect to waste determinations performed pursuant to the NDAA, NRC staff notes it is statutorily limited to assessing compliance with the performance objectives of Subpart C of 10 CFR Part 61. Similarly, NRC staff expects to assess whether waste at Hanford or West Valley will meet the performance objectives of Subpart C, or comparable performance objectives, because Subpart C is specified in the relevant criteria (see Sections 2.2 and 2.3). As noted in the guidance, staff is directed to consider other parts of the regulation, as appropriate, to provide additional context for certain review areas. For example, the guidance directs staff to consider disruptive processes listed in the siting requirements in 10 CFR 61.50 when assessing compliance with the site stability performance objective (10 CFR 61.44).

COMMENT: One commenter suggested that NRC and DOE have misconstrued the meaning of the term "waste incidental to reprocessing" and should "cease efforts to justify the incidental waste concept as anything other than a recently created fiction."

RESPONSE: NRC staff disagrees with the commenter's view that the concept of incidental waste is novel or that the historical development of this concept as described in the Background section of the Introduction is not supported by the cited documents. NRC for many years has employed criteria to determine that certain reprocessing wastes do not need to be disposed of in a geologic repository because they are "incidental" wastes; *see, e.g.,States of Washington and Oregon: Denial of Petition for Rulemaking*, 58 FR 12342, 12345; March 4, 1993.

COMMENT: One commenter expressed the view that the statement in the Introduction that DOE could use the criteria in DOE Order 435.1 to make an incidental waste determination is inaccurate because when the U. S. Court of Appeals for the Ninth Circuit vacated the decision of the U.S. District Court of Idaho [NRDC v. Abraham, 271 F.Supp.2d 1260 (D. Idaho 2003)], it did so only on the basis of ripeness and the criteria are in plain violation of the Nuclear Waste Policy Act of 1982. *See NRDC v. Abraham*, 388 F.3d 701 (9th Cir. 2004).

RESPONSE: NRC recognizes that the commenter disagrees with the views of the Government in the above-referenced litigation on the legality of making incidental waste decisions under DOE Order 435.1. However, the Ninth Circuit's action vacating the District Court's decision means that there is no court decision barring DOE from using the criteria in DOE Order 435.1 to make an incidental waste determination.

COMMENT: One commenter expressed the view that the guidance in the draft document cannot be applied legally to make WIR waste determinations at the West Valley site because it appears that any use of WIR determinations at that site would violate both the Nuclear Waste Policy Act of 1982 (NWPA) and the West Valley Demonstration Project Act of 1980 (WVDPA). The commenter stated that, in both of these laws, Congress created waste classification systems that define high-level waste (HLW) in a way that cannot be overruled by NRC. The commenter expressed the concern that NRC was providing DOE with "de facto" authority to dispose of its wastes onsite at the eventual expense of New York. Another commenter has serious doubts about whether the WIR criteria can be used at West Valley.

RESPONSE: The NRC does not agree that the definitions of "high-level waste" in either the NWPA or the WVDPA conflict with the ability to make WIR waste determinations. The NRC has explained its view that Congress' definitions of HLW in those Acts incorporated the understanding of the Atomic Energy Commission and of the NRC that HLW does not include incidental waste, *see, e.g., Proposed Rule: Disposal of Radioactive Waste*, 53 FR 17709 (May 18, 1988). Thus, NRC's participation in a WIR determination does not "overrule" the statutory definitions of HLW. The WVDPA requires NRC to prescribe decontamination and decommissioning criteria to be used by DOE at West Valley. In doing so, NRC established the criteria to be used by DOE for WIR determinations at West Valley, *see Decommissioning Criteria for the West Valley Demonstration Project (M-32) at the West Valley Site; Final Policy Statement*, 67 FR 5003, 5011-5012 (Feb. 1, 2002), but DOE relies on its own statutory authority in making WIR determinations.

COMMENT: The same commenter objected to the suggestion in the guidance that DOE Order 435.1 may authorize WIR determinations for waste sent offsite from the West Valley site for the same reasons as expressed in the previous comment.

RESPONSE: DOE's ability to make WIR determinations does not conflict with the NWPA or the WVDPA for the same reasons as given in the previous response.

COMMENT: One commenter suggested that the procedures and requirements set forth in the guidance document should be codified into regulations and applied to all sites where WIR determinations are permitted.

RESPONSE: The guidance document does not contain requirements for NRC or DOE; rather, it provides guidance to the NRC staff for evaluating DOE-developed waste determinations for SRS, INL, Hanford, and West Valley. As explained in the Introduction, the criteria used by DOE in making WIR determinations is dependent upon the relevant statutory regime governing the site. NRC does not have regulatory authority over DOE with respect to its WIR determinations and so would not be able to adopt regulations applying to DOE in this matter.

COMMENT: One commenter suggested that "Considering the factors for compliance with the performance objectives when reviewing the West Valley determination would seem appropriate if it is the NRC intent to rely on it when the license is reinstated. In this way the NRC staffs developing the factors are fully aware of the determination and its basis."

RESPONSE: The NRC staff agrees that any factors important to compliance with the incidental waste criteria may need to be considered during the analyses of any proposed West Valley license modification or termination. At the time of this writing, DOE has not informed the NRC if or how DOE plans to submit waste determinations for any onsite incidental waste disposal at West Valley (e.g., as separate analyses or as part of a decommissioning plan). Therefore, the method of documenting any incidental waste analyses will necessarily be decided at a later date. Section 9 of the guidance document was clarified as a result of this comment.

Advisory Committee on Nuclear Waste (ACNW) Comments

COMMENT: The ACNW noted that radionuclide removal efficiency is not a meaningful measure of risk. The Committee recommended that the review guidance concerning DOE's decision that radionuclides have been removed to the maximum extent practical should focus on reducing risk from the residual radionuclide inventory and the costs (e.g., financial costs, potential increases in worker risk) of achieving risk reduction.

RESPONSE: NRC staff agrees with the Committee's view that radionuclide removal efficiency is not a direct measure of risk. In general, no statement of removal efficiency alone is expected to provide a sufficient basis to conclude that radionuclides have been removed to the maximum extent practical. Text has been added to Section 3.3 to emphasize this point.

However, there are several reasons that NRC staff may need to review radionuclide removal efficiencies. In general, the term "removal efficiency" may refer to fraction of waste removed from a contaminated structure or the fraction of a radionuclide removed from a waste stream (e.g., by chemical removal methods). Thus, radionuclide removal efficiencies often are directly related to radionuclide inventory, which typically has a significant effect on the potential doses from a disposal facility. If radionuclide removal is not complete when NRC staff reviews DOE's draft waste determination, NRC staff may evaluate DOE's projected removal efficiencies and associated uncertainties as one aspect of its evaluation of radionuclide inventory. In addition, efficiencies may be used as one part of the demonstration that radionuclides have been removed to the maximum extent practical or the demonstration that potential doses to members of the public will be ALARA (as required by 10 CFR 61.41). For example, projected removal efficiencies may be used as a factor to compare alternate radionuclide removal methods or technologies. In addition, DOE may cite declining removal efficiency as a reason for ceasing removal activities [e.g., as in waste determinations performed for tank closure at INL and SRS (DOE, 2005a, b)]. If DOE reports declining removal efficiency to show a chosen technology has performed to the limit of its usefulness, reviewers must verify the removal efficiencies to evaluate DOE's basis for ceasing removal activities. The review areas and review procedures related to removal efficiency have been revised to clarify appropriate uses of removal efficiencies. In addition, new review procedures have been added so reviewers can evaluate whether DOE considered options to improve removal efficiencies.

NRC staff agrees with the Committee's comment that the review of DOE's decision that radionuclides have been removed to the maximum extent practical should focus on the potential reduction of risk from the residual radionuclide inventory and the costs of achieving risk reduction. NRC staff believes Section 3.4 of the guidance focuses on this comparison. The draft document also includes guidance for staff reviewing DOE's identification of highly radioactive (or key) radionuclides and DOE's selection of radionuclide removal technologies and methods. NRC staff believes that these areas of review are appropriate and has retained these sections of the guidance.

COMMENT: The ACNW recommended that a traditional approach to collective dose should not be used to evaluate DOE's decision that highly radioactive (or key) radionuclides have been removed to the maximum extent practical and suggested that the dose to the reasonably maximally exposed individual (RMEI) or average member of the critical group should be used instead.

RESPONSE: Section 3.4.1 of the draft guidance describes reasons that the $2,000 per person-rem factor that NRC uses to convert collective dose to a dollar value in some contexts (e.g., regulatory analyses and ALARA analyses for license termination) may not be applicable to DOE waste determinations. As described in the draft guidance, it appears to be more appropriate to compare the costs and benefits of additional radionuclide removal to the costs and benefits of other similar DOE activities.

In its comments, the Committee referenced a previous letter to the Commission in which the Committee expressed its view that collective dose has little value as an absolute measure of risk (ACNW, 2005). In that letter, the Committee also stated that collective dose is useful for comparing different management options (e.g., in ALARA analyses). NRC staff agrees with this position (NRC, 2005) and believes analyses of the practicality of additional radionuclide removal are similar to ALARA analyses in that the essential decision is whether more should be done to reduce doses once the applicable dose limits are met. Furthermore, NRC staff believes that DOE should choose how to quantify benefits of additional radionuclide removal and that the review guidance should prepare reviewers to evaluate a variety of potential cost-benefit comparisons. Thus references to collective dose calculations have not been removed from Section 3.4.1, but the text has been revised to shift the emphasis from traditional collective dose metrics to comparisons based on doses to individuals (e.g., the RMEI or average member of the critical group). Additional detail has been added to the discussion of potential difficulties associated with quantifying benefits in terms of collective dose in analyses supporting waste determinations.

COMMENT: The ACNW made several specific recommendations about factors that should be considered in NRC's review of DOE's selection of radionuclide removal technologies. In addition, the Committee recommended that NRC staff should monitor the status of less mature radionuclide removal technologies for potential future consideration.

RESPONSE: NRC staff agrees with the Committee's recommendation that the review of DOE's radionuclide removal technology selections should focus on systematic consideration of the integrated cost, technology maturity, and extent to which additional radionuclides might be removed by less mature technologies. Section 3.3 of the guidance focuses on these considerations. The draft guidance directs reviewers to verify that DOE considered an appropriate range of technologies, including technologies developed across the DOE complex. In response to the Committee's comments, text has been added to Section 3.3 to direct reviewers to evaluate whether DOE considered information from international and industrial sources.

The Committee also noted that application of a suite of technologies may be more successful than repeated application of a single technology and recommended that the review of technology selection also include consideration of any potentially beneficial sequencing of and synergism among candidate technologies. NRC staff agrees with this position and has added similar text to the review areas and review procedures of Section 3.3. In addition, new review procedures have been added to direct reviewers to verify that DOE considered whether modifying its approach (e.g., redirecting mixing jets or pumps) or applying an alternate technology could practically improve removal.

NRC staff agrees with the Committee's recommendation that NRC staff should monitor the status of less mature radionuclide removal technologies for potential future consideration and have noted this suggestion in Section 3.3. However, as the Committee recommended, technology selections in waste determinations are expected to focus on relatively mature technologies. One possible exception may be waste determinations that DOE submits several years before beginning radionuclide removal, in which case NRC staff may make an extra effort to learn about less mature technologies that may be mature by the time DOE begins removal activities. However, in most cases, reviewers are not expected to focus on less mature technologies during the review of a waste determination. Therefore, NRC staff have not added any review procedures specifically related to less mature technologies to the review guidance.

COMMENT: The ACNW commented that the guidance should indicate that there may be circumstances where blending of different waste classes may be appropriate

RESPONSE: Additional emphasis has been added to the existing text to indicate that blending of waste streams should not be undertaken solely for the purpose of waste classification.

COMMENT: The ACNW commented extensively on the concentration averaging guidance, and they recommended developing a more risk-informed approach consistent with the specific characteristics of the waste, disposal site, and method of non-HLW waste disposal.

RESPONSE: The concentration averaging guidance has been revised to add a third approach, labeled Category 3. The new approach is more risk informed in that it allows consideration of the appropriate intruder scenario when determining waste classification. Specific example averaging expressions for different scenarios have been developed for staff to use as benchmarks when reviewing waste classification calculations. However, the Category 3 approach entails a greater burden in completing the classification calculations and in staff review effort of those calculations. Note that a number of aspects specific to the disposal of commercial LLW were considered when the concentration limits and classification system found in 10 CFR 61.55 was developed. Therefore, development of concentration averaging for non-HLW is significantly more complicated than simply taking the concentration limits in 10 CFR 61.55 and correcting for a new dilution volume consistent with a new intruder scenario.

COMMENT: The ACNW commented that the guidance should not indicate a need for more stringent disposal requirements for greater than Class C waste because of large variation in the natural system capabilities to isolate waste from site to site.

RESPONSE: The guidance about the stringency of disposal requirements for greater than Class C waste comes from the agency position in 10 CFR Part 61. The staff agrees that some sites may have natural system characteristics that result in a higher likelihood for waste isolation. However, the waste classification system (concentration limits), is more directly related to intruder scenarios that are not strongly influenced by natural system characteristics, but are more strongly influenced by the depth and distribution of waste and especially the disposal requirements.

COMMENT: The ACNW recommended deleting the guidance that the intruder scenario should not be probability weighted.

RESPONSE: The staff disagrees that the guidance should be deleted. The likelihood of an intruder event occurring (of a future person or persons intruding into a commercial waste facility) is highly uncertain. The regulatory framework for the commercial disposal of LLW considered that it would be unlikely that someone would disrupt waste in the future, but not impossible. Therefore 10 CFR 61.42 was developed (based on reasonably conservative scenarios that were not highly speculative) to ensure, in the event someone intruded into the waste disposal facility, that public health and safety would be protected. The primary disruptive scenarios that were envisioned were drilling to recover water or other resources or construction of a residence.

To allow the probability weighting of intruder scenarios could appear to be more risk informed (e.g., risk = probability times consequence). However, the calculation of risk needs to include the impact of uncertainty. Even over the last 200 years, society has changed dramatically. Many natural processes can be forecast with some degree of confidence over long periods of time. However, human activities are more uncertain because they are nonstationary.

The intruder dose limit that is commonly applied (though it is not specified in the regulation) is 5 mSv [500 mrem]. The consequence value is 20 times larger (and therefore the implied probability of the scenario occurring is 20 times smaller) for the intruder compared to the offsite member of the public. In addition, the consequence value of 5 mSv [500 mrem] is generally comparable to the average level of exposure to a member of the public from other sources. The current approach ensures that the maximum dose a member of the public would receive is 5 mSv [500 mrem] in the unlikely event that the intruder scenario would occur.

Note that waste depth is a primary determinant to the likelihood that waste may be disrupted, and waste depth is considered when defining the appropriate intruder scenario.

COMMENT: The ACNW commented on a number of aspects regarding probabilistic and deterministic approaches to performance assessment.

RESPONSE: The guidance document clearly indicates the preference for a probabilistic performance assessment, including discussion in Sections 4.0 and 4.4.1.1 on the advantages of the approach. Section 4.5 provides review procedures specific to assessing sensitivity and

C–27

uncertainty when a deterministic performance assessment has been completed. Specific review procedures are provided for data uncertainty, model uncertainty, and model development for probabilistic and deterministic analysis. Waste determinations may be performed for many different types of waste and systems. The guidance document adequately directs the staff assessment of both types of analyses and helps staff determine when a deterministic analysis may be insufficient. If there are a number of key uncertainties, a factorial analysis of parameter and model uncertainty with a deterministic model should be able to assess the impacts of uncertainty, albeit in a somewhat more cumbersome way, just as a probabilistic assessment would. Model support is a much more direct determinant of performance assessment adequacy than the technique used to perform the assessment.

COMMENT: The ACNW commented that the information supporting the performance assessment should be robust and that postclosure monitoring should not be relied upon as a substitute for inadequate information.

RESPONSE: The intent of the guidance document is to ensure that the performance assessment is a robust assessment of system performance. However, many uncertainties are associated with most performance assessments, and monitoring is a mechanism to manage those uncertainties. Monitoring is also a way to evaluate new information that may confirm or refute previous information. Assumptions are part of any analysis, including performance assessments. Monitoring can be used to verify assumptions; however, the staff agrees that the performance assessment must be adequately supported to demonstrate compliance with the performance objectives. The main elements within Section 4 (system description, data sufficiency, data uncertainty, model uncertainty, and model support) were specifically developed to ensure a technically adequate performance assessment. However, the staff also recognizes that some uncertainties will be unable to be reduced when decisions are to be made. It is not practical to require an assumption-free performance assessment. It is reasonable to expect that key assumptions have adequate technical basis and that the assumptions can be verified during postclosure monitoring. Monitoring is not to be used as a substitute for inadequate information, but rather a confirmation of the previous determination of adequacy considering uncertainty. The guidance has been clarified in Sections 4 and 10.

COMMENT: The ACNW commented on a number of aspects related to the dosimetry used in the performance assessment.

RESPONSE: The guidance document advocates the use of dosimetry consistent with 10 CFR Part 20. This ensures that compliance calculations between 10 CFR 61.41 and 10 CFR 61.43 remain consistent, as is discussed in NUREG–1573, "A Performance Assessment Methodology for Low-Level Radioactive Waste Disposal Facilities."

The proposed rule for 10 CFR Part 63, "Disposal of High-Level Radioactive Wastes in a Proposed Geological Repository at Yucca Mountain, Nevada; Proposed Rule," states "As a matter of policy, NRC considers 0.25 mSv [25 mrem] Total Effective Dose Equivalent (TEDE) as the appropriate dose limit within the range of potential doses represented by the current 10 CFR 72.104 limit of 0.25 mSv [25 mrem] (whole body), 0.75 mSv [75 mrem] (thyroid dose), and 0.25 mSv [25 mrem] (to any other critical organ)." (64 FR 8644)

As 10 CFR 61.41 has the same standard as 10 CFR 72.104, this policy is applicable and therefore, Waste Incidental to Reprocessing (WIR) determinations will use TEDE, without specific consideration of individual organ doses. In addition, the guidance specifies that intruder calculations should also be based on 5 mSv [500 mrem] TEDE calculations, without specific consideration of individual organ doses, to be consistent with the dosimetry used for 10 CFR 61.41 and 10 CFR 61.43. Because of the tissue-weighting factors and the magnitude of the TEDE limit, specific organ dose limits are not necessary to protect from deterministic effects.

A current NRC policy allows licensees to use these systems on a case-by-case basis if they request them. While not completely identical in detail, NRC would equate 0.25 mSv [25 mrem] effective dose as the same as 0.25 mSv [25 mrem] total effective dose equivalent. In NUREG–1757, Volume 2, Appendix I.5.3.2, NRC addresses the issue for long-term performance assessment within the scope of decommissioning. The guidance states

> "Licensees may request an exemption from Part 20 to use the latest dose conversion factors (e.g., ICRP 72). Scenarios and critical group assumptions should be revisited to look at age-based considerations. Licensees may not 'pick and choose' dosimetry methods for radionuclides (e.g., Federal Guidance Report No. 11 for six radionuclides and current International dose conversion factors for three radionuclides)."

DOE is not a licensee and would not require an exemption to use the alternate dosimetry. However, the guidance for WIR would be consistent in that DOE should use the alternate dosimetry for all parts of the analysis (10 CFR 61.41, 61.43, and intruder analyses) and for all radionuclides. An exception may be for direct exposures, which were not updated for ICRP–72 and therefore would need to use the FGR-13 values. Additionally, staff would consider whether age-specific scenarios or exposure groups could result in higher doses.

References

Advisory Committee on Nuclear Waste (ACNW). "Comments on USNRC Staff Recommendation of (sic) the Use of Collective Dose." Letter from M. Ryan to N. Diaz, NRC. September 2005.

National Academy of Sciences (NAS). "Tank Waste Retrieval, Processing, and On-Site Disposal at Three Department of Energy Sites." 2006.

International Commission on Radiological Protection (ICRP). "ICRP Publication 60: 1990 Recommendations of the International Commission on Radiological Protection Annals of the ICRP Volume 21/1-3." April 1991.

U.S. Nuclear Regulatory Commission (NRC). "Response to Letter from the Advisory Committee on Nuclear Waste, Dated September 30, 2005, on 'Comments on USNRC Staff Recommendation of (sic) the Use of Collective Dose.'" Letter from N. Diaz to M. Ryan, ACNW. October 2005.

———. "Reassessment of NRC's Dollar Per Person-Rem Conversion Factor Policy." NUREG–1530. December 1995.

———. "A Performance Assessment Methodology for Low-Level Radioactive Waste disposal Facilities." NUREG–1573. October 2000.

———. "Decommissioning Criteria for the West Valley Demonstration Project at the west Valley Site; Final Policy Statement." *Federal Register*, 67 FR 5009. February 2002.

———. "Consolidated Decommissioning Guidance." NUREG–1757, Vol.2. September 2006.

U.S. Department of Energy (DOE). "Draft Section 3116 Determination Idaho Nuclear Technology and Engineering Center Tank Farm Facility." DOE/NE–ID–11226. DOE, Idaho Operations Office. 2005a.

U.S. Department of Energy (DOE). "Draft Section 3116 Determination for Closure of Tank 19 and Tank 18 at the Savannah River Site." DOE–WD–2005–002. DOE-Savannah River. 2005b.

NRC FORM 335
(9-2004)
NRCMD 3.7

U.S. NUCLEAR REGULATORY COMMISSION

BIBLIOGRAPHIC DATA SHEET

(See instructions on the reverse)

1. REPORT NUMBER
(Assigned by NRC, Add Vol., Supp., Rev., and Addendum Numbers, if any.)

NUREG-1854

2. TITLE AND SUBTITLE

NRC Staff Guidance for Activities Related to U.S. Department of Energy Waste Determinations

3. DATE REPORT PUBLISHED

MONTH	YEAR
August	2007

4. FIN OR GRANT NUMBER

5. AUTHOR(S)

NRC: H. Arlt, A. Bradford, N. Devaser, D. Esh, M. Fuller, A. Ridge
CNWRA: B. Brient, P. LaPlante, P. Mackin, E. Pearcy, D. Turner, J. Winterle

6. TYPE OF REPORT

7. PERIOD COVERED *(Inclusive Dates)*

8. PERFORMING ORGANIZATION - NAME AND ADDRESS *(If NRC, provide Division, Office or Region, U.S. Nuclear Regulatory Commission, and mailing address; if contractor, provide name and mailing address.)*

Division of Waste Management and Environmental Protection
Office of Federal and State Materials and Environmental Management Programs
U.S. Nuclear Regulatory Commission
Washington, D.C. 20555-0001

Center of Nuclear Waste Regulatory Analysis
Southwest Research Institue
6220 Culebra Road
San Antonio, TX 78238-5166

9. SPONSORING ORGANIZATION - NAME AND ADDRESS *(If NRC, type "Same as above"; if contractor, provide NRC Division, Office or Region, U.S. Nuclear Regulatory Commission, and mailing address.)*

Same NRC address as above

10. SUPPLEMENTARY NOTES

PROJ0734, PROJ0735, PROJ0736, POOM-32

11. ABSTRACT *(200 words or less)*

This document provides guidance to the staff of the U.S. Nuclear Regulatory Commission (NRC) for conducting activities related to waste determinations. Waste determinations are evaluations performed by the U.S. Department of Energy and are used to assess whether certain wastes resulting from the reprocessing of spent nuclear fuel can be considered low-level waste and managed accordingly. This guidance document applies to NRC activities conducted for the Savannah River Site (SRS) in South Carolina and the Idaho National Laboratory (INL) in Idaho pursuant to the Ronald W. Reagan National Defense Authorization Act for Fiscal Year 2005 (NDAA). The document also applies to the Hanford site in Washington, and the West Valley site in New York. The guidance document provides information regarding the background and history of waste determinations, the different applicabe criteria and how they are applied and evaluated, and NRC's monitoring activities that will be performed at SRS and INL in accordance with the NDAA.

12. KEY WORDS/DESCRIPTORS *(List words or phrases that will assist researchers in locating the report.)*

Incidental waste. High-level waste tanks. Waste determinations. WIR. Waste incidental to reprocessing. Savannah River site. Idaho National Laboratory. Hanford. West Valley.

13. AVAILABILITY STATEMENT

unlimited

14. SECURITY CLASSIFICATION

(This Page)

unclassified

(This Report)

unclassified

15. NUMBER OF PAGES

16. PRICE

NRC FORM 335 (9-2004)

PRINTED ON RECYCLED PAPER